THE BANNER BRAND

SMALL BUSINESS SUCCESS
COMES FROM A BANNER BRAND
- BUILD IT ON A BUDGET

Mark A. Cenicola

THE BANNER BRAND

SMALL BUSINESS SUCCESS COMES FROM A BANNER BRAND - BUILD IT ON A BUDGET

Mark A. Cenicola

BannerView Press

The Banner Brand: Small Business Success Comes from a Banner Brand - Build it on a Budget
First Edition
Copyright © 2011 by Mark Cenicola

Published by BannerView Press
  6348 S. Rainbow Blvd. Ste 110
  Las Vegas, NV 89118
  702-312-9444
  http://www.bannerviewpress.com

Cover art by BannerView.com

Editing and layout by Circumference Communication
  www.CircumferenceCommunication.com

ISBN: 978-0-9829085-2-5

Printed in the United States of America

ACKNOWLEDGMENTS

There are a few people I'd like to acknowledge that really made this book possible, starting with my wife Elizabeth. Working full time, she held down the fort at home while I worked even longer hours, and networked like crazy. She put up with me partially paying attention at home as inspiration for this book came and went. I want to thank her for not getting too upset for not taking her with me as I traveled to various places to conduct business. She did all this while carrying our baby girl Isabella.

Joseph Stanley and Jeff Helvin, both of whom I've known both for nearly 20 years, have helped me build an incredible company where I am free to focus my energies where necessary and can entrust the operations while I was away writing, speaking, networking, and traveling.

Thanks goes to Lisa Schonsheck, my Executive Assistant who handles 98% of my customer relationships, keeping them happy so that I'm needed very little.

Thanks goes to Robert & Samantha for treating me as if I was their biological father and being patient while I worked to support our family.

Thanks goes to Jonathan Peters for inspiring me to publish this book, and for helping me realize that authorship was not only obtainable, but also a great way to build credibility.

Thanks goes to Spencer Overman for designing an awesome book cover - may the force be with you for it!

Thanks goes to everyone I didn't mention specifically that has had influence in my life. You know who you are, so go ahead and take credit because it's well deserved!

Finally, thanks goes to Isabella who was just developing into a human being while most of this book was being written. What a most joyous experience it has been to see you come into this world and give me the gift of true love at first sight!

TABLE OF CONTENTS

INTRODUCTION

The Banner Brand will give you golden nuggets of information you need to make your business a success. Whether you are starting a new business, turning around an old business, growing a small business, looking for inspiration to reenergize a large company, or simply trying to find a reason why you are still slaving away at your current business, job, or whatever else you do, this book will provide you with ideas that will contribute to your success.

While I don't expect you to hang on to every word or piece of advice, it's quite reasonable to think that in the process of going through each chapter, you'll find a banner nugget that is worth the money you spent purchasing and the time you spent reading this book.

As you walk through the journey I've taken to get where I'm at today, you'll realize that it's never who you know or what you know, but rather who you can get to stand behind and believe in your brand.

The power and value a Banner Brand can generate can't be measured on paper or reported on a set of financial statements, but it's no doubt a force in business. Building a successful brand will make everything else you do in business seem like a stroke of good luck.

Little time will be wasted with a lengthy introduction to *The Banner Brand* because you are a busy person who wants to get right to the meat of building your own Banner Brand. Go ahead, sit back, and begin your pursuit of Banner Nuggets. Implement those nuggets into your business. And with patience and perseverance, watch your success increase.

Then, spread the good word by sending a copy of *The Banner Brand* to your clients, vendors, and associates whom will have you to thank for helping them be successful in their business endeavors.

Finally, visit TheBannerBrand.com to share your success stories, ideas, and what you've done to build your Banner Brand. We can share Banner Nuggets with our readers who may use your insights and experiences to take their businesses to a new level. They will have you to thank for their success!

CHAPTER 1
SHIT FOR BRANDS

*T*he first two entries for the definition of the word "brand" (noun) at Dictionary.com are:

1. Kind, grade, or make, as indicated by a stamp, trademark, or the like

2. A mark made by burning or otherwise, to indicate kind, grade, make, ownership, etc.

The first definition, in the commercial sense of brand, corresponds with a company's logo. A company's brand is usually represented by a logo. It signifies ownership and stands for the quality a company represents.

The second definition, while more literally meaning the mark being left behind as a result of burning or engraving (such as with cattle), can also be seen as the impression ingrained in the minds of the people who experience a company's brand.

Brands come in many different representations that make good, bad or simply no impressions. Building, promoting and maintaining a brand that represents a certain quality and is recognizable is a key ingredient to success in any business.

When people think of brands, usually what comes to mind are huge Fortune 500 companies like Coca-Cola, IBM, Microsoft, GE, Nokia, Google, Toyota, McDonalds, Apple, etc. Why are these companies the top companies in their industries? Is it because they have great technology, addictive food, great cars, or tasty soft drinks? No! It's because they have tremendous brand value that keeps them at the top in consumers' minds.

Do you recall the saying, "No one ever got fired for buying IBM?" There are certainly companies that make better tasting food, faster cars, healthier drinks and more advanced technology, but the problem is that those companies simply can't compete with the power of these brand superstars.

So what chance does a small company have in competing in a business world with such large and powerful brands? I'll share some good news, bad news, and even more good news.

Starting with the good news, the platform in which small businesses operate is significantly different than large companies. Small business customers are different than the customers who purchase from large businesses since they are looking for something different, unique, or they simply want personalized attention that the large businesses can't or do not offer.

The bad news is that small businesses vastly outnumber large businesses, which means there is more competition.

In order to be a successful small business, you need the power of a great brand. Many small business owners overlook branding as a key to success because they believe branding is left only to the large and powerful businesses. That couldn't be farther from the truth.

The adjective definition of the word "banner," according to Dictionary.com, is "leading or foremost." To be successful, it's not enough to have just any brand. You must position your brand as *THE* Banner Brand. You need the brand that stands out as the leading or foremost choice. With a proper plan in place, your small business can utilize affordable and guerilla tactics that can successfully position your brand as the brand of choice to your set of customers.

The even more good news is you can start developing a Banner Brand by looking at both good and bad examples from the the largest companies. They've spent millions of dollars experimenting with their brands, so learn from their successes and avoid their mistakes.

In our first example, let's look at a terrible branding decision by Coca Cola®. In the mid 1980s, Coca-Cola made the decision to disrupt their sales of a highly successful soft drink loved by a devoted set of customers. Like many marketing people who get bored, they made a decision to introduce a "New Coke" and bet the house on it.

We all know how that turned out. Coca-Cola quickly changed back to their old brand. This venture into a new type of soft drink might have ben successful if they introduced it with a new name and leveraged the old brand. They could have called it "XYZ Cola," brought to you by the makers of Coca-Cola. They wouldn't have alienated their customers by changing what their brand stood for, and they could have placed the strength of their existing brand behind this new product. Or, they could have built a new brand from scratch completely separate from their highly successful Coke product. Any of these options would have had little risk of damaging their Banner Brand.

Another more recent example of a terrible branding idea is Dominos. Dominos had widely been known for getting pizza to hungry people very quickly as backed up by their slogan of "pizza in 30 minutes or less." They all but abandoned that concept and simply became another pizza company. They still were able to maintain a great brand since customers stayed loyal to a pizza recipe they came to know and love.

However, Dominos has completely changed their pizza recipe in an attempt to reinvent themselves. By doing so, they have now alienated those customers who liked the original pizza recipe. Therefore, they are no longer known for delivering pizza in 30 minutes or less, and they don't have a real reason for past customers to keep buying from them. It's true they may get new customers who like the new recipe, but a company who's had a tremendous brand for many years, and known and trusted by legions of customers, has abandoned those customers. They've changed what their brand stands for and the product behind the brand. If they wanted to make such a big change, why not start a new company, with a new brand that catered to the type of pizza they now wanted to make?

A company with a great branding strategy and Banner Brand is Apple. The company continues to make products that nobody thought they needed, and Apple sells a ton of them.

Apple is a premium brand that represents hip and cool. People don't own Apple products because they need them; they own them because they want them. When Apple releases a product, it's seen in the eyes of consumers as a luxury. However, it's the kind of luxury that's still within reach of many consumers. People love affordable luxury, which keeps them coming back for more.

While Apple has a great brand now, there was a time when things weren't so rosy. Prior to Steve Jobs coming back to Apple, management made the decision to open up the Apple platform to clone manufacturers. This made Apple products no longer luxury items, but over-priced commodities. It was a recipe for disaster which Jobs quickly remedied upon his return as CEO.

Our last example is a company with a Banner Brand, but I believe they are wastefully spending more than they need to preserve that Banner Brand. That company is Pepsi.

When I was growing up, the biggest decision I had to make was Pepsi or Coke. I sometimes drank Pepsi since my mother was a Pepsi drinker, but sometimes I drank Coke since my father was a Coke drinker. Ours was a dual soft drink branded household, and we all got along!

As I got older, I grew to choose Coke over Pepsi, not so much because of taste, but because Pepsi seems to change their can design every couple of years. I no longer have the same attachment to Pepsi because it doesn't look like the same Pepsi I had when I was a kid. Brand inconsistency turned me to their competitor, and it had nothing to do with a change in product quality.

What's the purpose of investing in a brand that's going to change year after year? It's a waste of money and confuses customers. In this instance, Pepsi looses loyal customers when they change brand design. Yes, from time to time they introduce retro cans that go back to older versions of their brand, which may attract people like me who identify with those old design. And, truthfully, I do find myself wanting to drink a Pepsi when I see the retro cans. But they no longer have a loyal customer in me.

It's okay for brands to evolve, but changing your brand year after year or every few years is not an efficient way to spend marketing dollars. Imagine all the money Pepsi would have saved if they stayed consistent with their branding.

What you can learn from these examples is that consistency in branding will far outweigh any advertising you spend. How many companies drop tons of money into a branding campaign, but then abandon the message when some newly hired marketing guru comes up with a great idea? How about spending NO money on advertising until you have a brand that tells a story you're proud of? As a small business owner, you have limited resources which means you must use what you have wisely.

At Banner View, we did what most small businesses will do. We made all the classic branding mistakes. It took us many years to realize this as we changed our branding from marketing campaign to marketing campaign. There was no consistency from one brand to the next. Our reasoning? We wanted to generate fresh leads. If one campaign didn't generate leads we moved onto the next.

It's tempting to abandon a marketing campaign when you don't get an immediate response. But switching from brand to brand is a terrible idea since any marketing campaign takes time. We didn't understand this reality. By not developing what we wanted our brand to represent and by failing stay consistent, we wasted a lot of time and money.

In fact, in the early years of my company, we had a brand identity crisis. My ego was such that I wanted my name to be front and foremost. Therefore, we chose Cenicola-Helvin Enterprises (CHE for short) as the name of our company. We had a number of niche website properties that all were operated under the CHE umbrella. However, from the first year in business forward, 99% of revenues were generated under the BannerView.com brand where we offered services to build, promote, and maintain websites for the small business market. The CHE name was long, but when people finally understood what I was telling them, and matched the name on the card with my name, it was always a slight ego boost when people said "Oh, that's your name in the company name."

Well, after many years, I realized that anyone and everyone is CEO of their own company; it costs nearly nothing to start a business and provide yourself with a CEO title. And after nine years of introducing myself as, "Mark Cenicola, with Cenicola-Helvin Enterprises. We provide e-business solutions under the BannerView.com brand," not only did I get tired of having to explain so much, I was also tired of people not understanding any part of what I said.

We finally laid the brand identity crisis to rest (mostly) when we filed for a fictitious firm name so that we could officially do business as BannerView.com. The corporate structure continues to be Cenicola-Helvin Enterprises (which still feeds my ego), but the company name of BannerView.com allows me to pretend that I'm not in charge when needed and makes my introductions much simpler and easier to understand.

Now I'm Mark from BannerView.com, which builds great websites that earn recognition, respect and rewards. If that's all I get to say, which takes less than 8 seconds, than I'm happy because they know my name and company. I'll find a way to sell them later; for now, though, they've been introduced to the brand in a unique, yet concise, manner, which is a leg up on the competition.

After a year or two in business, the slogan "Give your Website A BannerView" came to mind. However, I never really bought into the slogan. I couldn't figure out what it meant; I just knew it sounded good.

Well, after nearly nine years in business, I reintroduced it to some staff in a marketing meeting, but decided to change "a" to "the," as in "Give Your Website *THE* BannerView"; it had more ego to it. I own *"The* BannerView," not just "A BannerView."

A simple word change allowed the slogan to come back from deep within, and now we have an applied a meaning to it. With the word *banner* meaning "leading" or "foremost," and *view* being pretty straight forward, we could now "Give Websites the BannerView." With a powerful meaning behind a nice sounding slogan, I was ready to be a brand ambassador.

Again, it took nearly nine years for this to happen, and another year to fully embrace the brand. If I had never spent a dime in marketing in the past, and saved all the money for a brand I truly believed in and embrace, I know I would be light years ahead financially. But hey, Warren Buffet said something to the effect of, "It's better to learn from other's mistakes than your own."

That's why you are still reading this book, because the money you spent to buy it is going to save you a ton of money in mistakes you would have wasted on advertising to non-loyal, price-driven customers.

Just think, you've come so much farther than most. Statistically, most of the people who buy this book won't read this far into it (few non-fiction books are ever read past page 16). Again, statistically, the rest of this book could be *Ipisis augiat, quam* and few would notice. While this book certainly looks great on the bookshelf, and I'm thankful to those who bought it (even though they'll never know I just thanked them), your investment will be paid off when you apply the information in the following pages to your business.

A recognizable brand is the first step to gaining a loyal customer base.

Customers who shop on price are not loyal—PERIOD —myself included. When I only want the lowest price, I buy what I need and never look back. Anyone coming to you because your price is lower than your competition will leave you as soon as someone else beats your price.

On the other hand, if your customers choose you because they know your brand, you have a step up on the competition. You can charge more, filter through the customers, have the opportunity make higher margins, charge more, and make more profits. If you happen to also have the best price, great, but remember, it's only temporary; someone can and will undercut your pricing. As long as you have a Banner Brand, loyal customers will not leave you when someone else offers a lower price.

Been there! We've tried price as a lure and got the kind of customers we didn't want. For us, price has not been a successful strategy for winning loyal business. We now have the ability to wait for the low price competitors to screw up. We have positioned our brand so that it is front of mind when potential customers beginning looking for alternatives after our competitors frustrate them.

In an world lacking trust and suffering from financial fraud and other schemes, each of us has narrowed the scope of how far we'll look for a product or service. I know I've narrowed the number of brands from which I buy. While I still compare price between my chosen brands, the narrower range has made me less of a price sensitive consumer.

As a company, BannerView.com has changed his focus from potential customers we didn't want to those we do, and it has made a world of difference. We now have the opportunity to go after the customers we truly want, those that will be loyal and profitable for us.

BannerView.com spent over $250,000 on advertising over nine years. While this isn't a lot in the grand scheme of things, it's certainly a lot for many small businesses. Plus, this number doesn't count the huge amount of time and employee costs that went along with building, revising, and revising again our website. Since we are in the business of building websites, our costs were soft costs, but it still means we wasted a lot of employee capital.

So where did all the hard money get spent? It was spent on postcard mailers, coupon mailers, business-to-business networking, phone directories even some TV commercials and radio.

The sad part is that we'd probably be in the same position today if we had only spent 20% of our advertising dollars. Why? Because our advertising didn't do anything to build our brand. For us, as for many small businesses, advertising was looked upon as a way to generate "hot" leads. Well, the majority of the leads we got were price shoppers. It seemed that the people who responded to the advertising may have needed the services we were offering, but they were shopping based on price. They hadn't heard of us before, so they only did business with us if we happened to have the best price at the time they were looking.

The few deals we got from our early advertising efforts turned out to be "cold" customers; they didn't stay with us for long. Loyalty was low and conversion of our leads was even lower because we didn't always have the lowest price.

So what worked the best? For us, the best customers came from business-to-business networking. The easy answer as to why networking consistently produces the best results is because we get in front of the best potential customers, build relationships with them, and, after they know us comfortably, they do business with us. Networking also produced a number of referrals that became loyal, long-time customers.

So how was the other 80% of our advertising dollars wasted? While we did get some loyal customers by chance, if we had spent those advertising dollars on joining and attending networking events, we'd be a better position today.

The advantages of networking is we have a clear mission when shaking people's hands: meet as many folks face to face as possible so they have a personal connection with our brand, then remind them every chance possible that we are here to serve their online business needs. We do this by adding all of the contacts from an event into our email newsletter database.

We send out newsletters like clockwork, providing consistent contact with those we meet at networking events. With that newsletter, we rarely, if ever, send out straight sales promotions. Instead, we have a newsletter rich in tips and advice for conducting business online so that our people will stay on our list and periodically read information relevant to them. The newsletter reinforces our brand with those we meet at networking events.

For instance, In one of my blog entries, I debated branding price and true value. The raw entry blog posting below shows how you need to get people to understand the true value of their purchase, and having a great brand certainly helps in that debate. It's the PC versus Mac debate.

As you will discover in the following blog entry, "PC" is not a single brand. PC is a collection of brands, but they've been grouped together as a single generic brand. This means that not only do manufacturers of PCs have to compete against the Mac brand, they are also waging an internal civil war against other PCs.

Let's say you go to the store to buy a new computer. The first decision you are going to make is PC or Mac. Let's say you choose PC. You have Dell, HP, eMachines, etc. to choose from. Does it really matter which you choose? You are just looking at the features of each product and comparing prices. Do you have any brand loyalty in making such a purchase decision once you've gone generic?

Branding involves making the decision to not be generic; it means placing value above price, and setting yourself apart.

The
BANNER
BLOG

The Age Old Debate - PC Versus Mac (Price versus Cost)

http://www.bannerview.com/banner-blogs/mark-cenicola/story.bv?storyid=1214

PC users and Mac users have debated since 1984 who's better, faster, cheaper, etc. However, very rarely do you hear the debate about productivity. Which platform will make you more productive? In my personal experience working between the Mac and PC platforms, I'm absolutely much more productive working in a Mac environment than a PC environment. I still hear it time and time again from Web developers who have come to work for BannerView.com that have previously been hard core PC users, that they are much more productive after making the switch to our Mac environment. Whether it's just the interface, speed of the machines, ease of use or absence of malware (viruses and spyware), there's certainly a productivity gap between the two platforms.

The main question is then, why don't more people use Macs?

The easy answer is price.

A new ad campaign by Microsoft (see the video at the end of this article) focuses on pricing differences between Macs and PCs. In a TV commercial, a woman named Lauren is given a $1,000 budget to find a 17" notebook computer. After visiting an Apple Store the only product in her budget was a 13" screen. She leaves with the comment, "I'm just not cool enough to be a Mac person." So she heads on over to Best Buy and purchases a $699 17" HP notebook. A 17" Macbook Pro would cost $2,799. Obviously, $2,100 is a big difference in price. Looking no further, a person would assume buying the $699 machine would be a no-brainer.

But let's take a look at the advantages that extra $2,100 offers. The 17" Macbook Pro, in comparison:

> has a 71% increase in screen resolution.
> has a 60% increase in memory speed.
> has more than a 100% increase in battery life.
> has a 27% increase in processor clock speed.
> has a LED screen display. The Advantages of LED
> > include brighter display and instant warmup.
> is 10% lighter.
> is 58% thinner.
> has Firewire 800 which is nearly double the speed of
> > USB ports.
> has a much faster graphics card.
> has a much faster wireless connection.
> has a much faster wired connection.
> has a multi gesture touch pad.
> has built in bluetooth.
> has a backlit keyboard.
> and, of course, has Mac OS X!

Assume you work 6 hours a day on your computer 5 days a week and that the advantages listed above make you *only* 15% more productive. If your time is worth a mere $15/hr., it would take 933 hours to make up the $2,100 difference in price. That translates into less than 32 weeks to recoup the price difference. If your time is worth $25/hr, you'd recoup your investment in 560 hours or less than 19 weeks. Now let's assume you bill at an hourly rate of $100/hr, you'd recoup your investment in 140 hours or less than 5 weeks! So, if you decided to purchase the Mac anyway because you were *cool* enough to do the math, you'd know that even if you financed the purchase over 1 or 2 years, you'd recoup your investment ever before spending $699 just from the increase in productivity! We are more productive which translates into savings for our customers.

Lauren is right, not everyone is cool enough to realize that there's a difference between price versus cost of a product or service. Convincing someone to make the switch from PC to Mac is a different story, but hopefully you get the point of the example anyway!

*** *The raw blog post contained a video so please view the online version to watch* ***

Winning a customer on price alone isn't necessarily a bad thing, and it's going to happen whether you like it or not. When this happens, you have the opportunity to brand after the fact.

Just because you closed a deal by offering the best price, doesn't mean you can't convert that price sensitive customer into a loyal customer. This route is certainly more difficult, but still achievable. You simply have to be more creative and prove that your customer made the right decision by going with your brand.

The following raw blog post talks about just one creative idea that we implemented to build brand loyalty from existing and potential customers.

The BANNER BLOG

Let the Tours Begin

http://www.bannerview.com/banner-blogs/mark-cenicola/story.bv?storyid=1290

Brian Mell, our Marketing Coordinator, put together a plan for us to offer tours of our facility to current and prospective clients so they could learn what BannerView.com does, what we can do for those looking to better conduct business online as well as see the operations behind what goes into developing a successful online presence. From that plan arose the BannerEdutainment Tour.

The tour idea was originally inspired by a tour Jeff Helvin and I took of Zappos.com's corporate headquarters in Henderson, NV. We learned some things to incorporate and not to incorporate into the BannerEdutainment Tour. While we had a good time on that tour and they certainly have a much larger facility, there really is no comparison between us and them - they sell shoes, we make it possible for companies to sell shoes should that be their need.

Our first tour was held August 28th and it seems the attendees had a great time and learned something in the process. While the tours are focused on helping us get more business, attendees learned some valuable tips and advice for doing business online, got to meet other business owners, got their photos taken with the BV Ride and enjoyed food and drink. We think that overall, the first tour was a success and look forward to the next two in September. The September 18th tour is fully booked, but as of this writing, there were a few spots open for the September 25th tour.

Enjoy the photos from our first group of tourists below and hope to see you at the next one!

*****Visit the original blog posting online to see the photos*****

When closing a cost-conscious prospective customer, you have just one brand impression—lowest (best) price. Hitting them with additional brand impressions, such as a physical tour of your office, an introduction to some of your other customers, or showing value-added benefits of being associated with your brand are key areas in which you can make a conversion.

The same holds true for customers who don't purchase solely based on price. The more opportunities you have to get your brand in front of those customers, especially via different mediums, the more comfortable your customers will be with their judgement that you are the right brand.

The following raw blog entry speaks to getting existing customers comfortable with your brand.

The BANNER BLOG

Intermingle Current Customers w/ Potential New Customers | The BannerEdutainment Tours Roll On!

http://www.bannerview.com/banner-blogs/mark-cenicola/story.bv?storyid=1298

As previously mentioned, we started inviting folks to visit the BannerView.com facility, meet our staff, have lunch on us and learn about the products and services available to help companies successfully navigate their businesses on-line. The BannerEdutainment Tour, as it's officially called, has so far gone over quite well. If you haven't taken the tour yet, there's another one being held on Friday, October 16, 2009. You can register right on our website.

While the tours are focused on introducing BannerView. com to those that may not be very familiar with us, we've had current customers attend that haven't previously seen all facets of our facility or know everything that we have to offer. When presented with the opportunity to have existing clients meet potential new clients, we jump at the chance as it's a way to demonstrate the loyalty we enjoy from many of our customers. It also serves as a great testimonial and in many cases, unsolicited at that. While we always get the warm and fuzzy feeling when customers compliment us on a job well done, there's nothing better that having others hear it first person. Posting testimonials on your website is one way to demonstrate positive feedback from current customers and video testimonials are certainly growing in popularity, however, there's simply no more powerful way to show it than in person.

As we continue to learn what works and what doesn't when it comes to our branding and marketing efforts, we are happy to share those experiences as we want our customers to learn from our successes and avoid our mistakes. In this case, the facility tours continue to create buzz and reflect positively on our brand. Should you have the opportunity to mix current customers with new potential customers in a first person environment, take the opportunity. Maybe it's time to start your own creative company tour. Just make sure we're invited. HINT: Email marketing is a powerful tool for that and we happen to have the perfect solution for you!

Please enjoy the photos from our two September 2009 BannerEdutainment Tours!

*****Visit the original entry online to see the photos*****

Your customers will be more loyal if they know other people who do business with you, or who at least know your brand.

People like to organize into groups with people who share similar ideas, values, and purchasing choices. You've probably heard of the herd mentality. Even if a customer's experience was negative, when they share that experience with others who have the same negative experience, they continue to be loyal. Unless one person in the herd moves, no one else will move, effectively giving you the opportunity to correct the negative situation before customers are lost.

This negative brand loyalty happens a lot in the telecom industry. It used to be that the large local monopoly was the only choice you had, but all of a sudden cable companies started offering a new technology called VOIP (Voice Over Internet Protocol). You may have had a bad experience with the telephone company, but so have all the other people you know. Even though you've had terrible service from the current company, you're afraid to switch to something "unknown"; therefore, you stick with the brand that you know.

Eventually someone in the herd will test the waters with the new brand. If they report a good experience, the rest of the herd will follow. But the switch won't happen all at once, and the new brand must have enough impressions with someone in the herd to build trust.

The point is even a brand that has had a negative impression still has the opportunity to keep loyal customers by improving the brand's perception. For instance, the company could offer better service before someone in the herd grazes new land.

In the following raw blog post, you'll see that I had just this experience with a telecommunications provider. While I didn't mention them by name, due to political reasons, not all bloggers are so forgiving.

The BANNER BLOG

Communication is Key in Customer Service

http://www.bannerview.com/banner-blogs/mark-cenicola/story.bv?storyid=1227

Recently, I ran into a situation with a company in which I've done business with for years. I was provided with a special offer for a free service upgrade, but it required a one year commitment. After the one year commitment they would begin to charge for the upgrade. It just so happens that the upgrade was on my wish list already so this incentive was a win-win proposition. I didn't mind sticking around for another year in the agreement and will continue to pay the new upgraded price after the free year is over which means more revenue for them.

Accepting that offer is where the fun started. The service was upgraded in a timely manner and I was happily taking advantage of it. Figuring that my bill wouldn't change because of the free upgrade I made the mistake of not paying attention to what I was being charged for a couple of months. Low and behold, I discovered that I was, in fact, being charged for the upgraded service. I figured this was a simple mistake and a quick call to customer service would resolve the issue. Customer service said they couldn't refund the previous months overpayment, but they would extend the free period for a year from when I notified them about the issue. I simply settled for this as the amount wasn't anything that couldn't be absorbed easily.

Another couple of months went by and again, the bills were for the full amount not accounting for the free upgrade. Now this started to frustrate me. This time, I decided to take a different approach and contact customer service via online chat as opposed to calling again on the phone. After the situation was explained, the customer service representative was happy to refund the previous months' overages and apply that amount as a credit on the next bill. In addition, they were to make sure that the offer would be enforced so the free upgrade would be reflected on future bills. Okay, so if you can't get what you want on the phone, the lesson is to try contacting them via a different method if possible? Well, sort of.

I made sure to be more diligent in looking for the next bill to make sure it had the corrected charges. It arrived via email, but there was no credit reflected on the bill, nor was the bill reflective of the free upgrade pricing. Now I was really upset. This time, I was going to try yet another method to get through to them and decided to send an email. I expressed my displeasure of dealing with this billing situation in the email. It's been a couple of days without a response so I was beginning to think that this is a lost cause. Checking my credit card statement, I found a surprise. Even though I haven't heard a response back to my email, I noticed that they did in fact apply the credit to my bill and only charged the amount, net of the credit. However, no where on my bill nor in my online account with the company does it mention anything of a credit. I just received a new bill, but it still reflects the incorrect pricing (without the free upgrade). I still have no idea where we stand on the billing situation or if they'll simply start charging my card the lower amount, but not reflect the charge on my official bill.

Prior to this, my view of the company was favorable. While I won't mention the name of the company, I'm certainly not eager to recommend their service anymore and I'm skeptical of upgrading to utilize additional services which I was seriously considering. All they had to do was communicate effectively with me and have their bills reflect the correct amount, including the amount they did credit back to me. Things happen, mistakes are made, but all could be forgiven if communication from the company was forthcoming.

Have you made a mistake on a customer's bill or been on the receiving end of an error? If not, you must be on a different planet as mistakes are commonly made. However, a little communication goes a long way toward resolving issues. Be forthcoming with what mistakes were made and how you are going to correct them and all will be well!

At BannerView.com, we've certainly made our share of mistakes from a service standpoint, and we haven't always done the best job promoting to our existing clients; however, since we started including all of our contacts, and not just prospective customers, in our branding efforts, we've seen an increase in loyalty from even price-conscious customers. Our efforts have also helped us identify those who don't have loyalty to us.

Employees, especially those who deal with customers, obviously play a critical role in portraying your brand's strengths and weaknesses. That's why it's crucial to make sure your employees understand and buy into the brand. After all, brand loyalty starts from within.

We did a poor job for many years portraying our brand and what it stood for within the company, mainly because we didn't know what our brand meant. Once we figured that out, it was a first priority to relay the brand's values to everyone on staff and get buy-in from them. Those who didn't or wouldn't buy into the brand will weed themselves out of the organization.

So where does brand creation start and end? It starts at the top. That means it starts with you: the business founder, owner, CEO, etc. It has to be your own brand. While you may involve others in the creation of such things as logo concepts, color variations, and creative design, the overall brand idea should be your creation; after all, you're more likely to have the passion necessary to make your brand a success, believe in the brand, and be the ambassador of that brand.

You can't expect others to buy into your brand if you don't believe in it yourself. Great brands aren't created from uncertainty; they are created from absolute certainty. Don't begin creating your brand if you don't know what you want. Don't settle for a brand you aren't completely sure about. And don't waste your time and money on having it developed if you aren't certain. So don't expect to leave the brand creation task completely to someone else. You must be actively involved in the brand creation process, provide constant feedback, and make absolute decisions.

Brand creation is a book all to it's own, and one that would be written by someone much more qualified than I, but no matter what your brand looks like, the lessons provided in these pages will be invaluable in promoting your brand.

The following raw entry from my blog further explains that we had to improve our brand as it progressed, as opposed to totally changing it. Since most people resist change, we knew we needed to take a weak brand and make it world class!

It's Not Change, It's Progress

http://www.bannerview.com/banner-blogs/mark-cenicola/story.bv?storyid=1196

How frustrating is it to learn something, have it down pat and then find out you've either been doing it all wrong or there's a new, better way to do things? People hate change and for good reason. It takes a long time to perfect habits, so to just change them suddenly is not always easy.

President Obama ran on the foundation of change, but other than the fact that the sitting President changed, politics in Washington will be politics as usual. The Democrats will argue their points and the Republicans will argue theirs with independents hoping someone will hear their ideas through the noise. Like a corporation or sports team making a change in management to turn things around, our country's change in administration should bring new ideas to the table to help progress our country forward. New laws will be passed, old laws will be undone and economic stimulus (stimuli) packages will be approved, but don't worry as none of this will dramatically change the way you live, your prosperity or relationships you've built. The good news is that you have the opportunity to adapt to these changes, putting yourself in a much better position to improve your situation both personally and professionally. Take control of what you can control, not fret about what you can't.

Over the years, we've gone through several progressions from making the decision to exclusively focus on our BannerView.com brand, to the many revisions we've made to our Web site. As mentioned in prior blog entries, I've hinted at things to come in 2009. We aren't changing our quality of service, the value of our offerings or even prices, but we are definitely adapting the way we sell and market our services. The failure of companies to adapt to a changing business landscape is usually the nail in the coffin. We look to this current economic climate as an opportunity that forces us to do better, be more innovative and focus on what makes us successful.

Over the next few days, you will see an improved BannerView.com Web site, a more focused marketing push and new technologies that will tie our service offerings more closely together. Some may call what we are doing as change, but I prefer to call it progress.

If you can develop a brand you truly believe in, you will reap very large rewards. Your thinking toward promoting that brand will be clear, and you won't waste money advertising a brand you don't believe in.

Take a good, hard look at your existing brand. After reading this book, you should have a number of ideas to help you promote it. However, you are only ready to promote it if you are ready to make progress with the brand you have now.

I'll close this chapter with a raw blog entry that speaks to the fact that you don't have to implement a bunch of things immediately to improve your brand's image. Small improvements over time, in any economy, can go a long way.

Small Improvements Can Make a Big Difference: Moving Forward Despite a Good and Bad Economy

http://www.bannerview.com/banner-blogs/mark-cenicola/story.bv?storyid=1300

The economy, the economy, the economy. Yes, it's important to know what's happening in the economy. However, the only information that's accurately reported about the economy is what's already happened. Predictions about the economy, no matter who the experts, are no better than that trusty crystal ball. While you can gain some insight into trends and areas to place your bets, the so called good news and the so called bad news isn't going to help make such predictions any more accurate.

What you need to decide is what is best for your business now despite the predictions about the economy. And what are the things that you are going to put in place so you can, not only survive, but thrive in any economy? By the time you wait for things to turn around, the train will have already left the station and you'll be in no better position then should you survive now. It seems to be one of those things where you are always catching up. So what can you do? Small things and lots of them. Small improvements over time can put you in a position to capitalize when the time is right. We are finding a number of businesses that have simply stopped doing anything and are focusing on what can be eliminated as opposed to what small things can be improved.

Take a look at your business, specifically your website. There are probably dozens of areas where you can make small incremental improvements and most of the time these small improvements are of little or no cost.

Here are small improvements, that over time, can lead to big results:

Get on a regular email marketing schedule
Email marketing is a very low cost branding opportunity for your business. Many companies already have email marketing capabilities in place, or if not, can add them for a very small investment. However, make sure your email newsletters are sent on a consistent schedule and have a consistent theme. Even when we've had to pull back on advertising and PR spending at BannerView.com, we've been dedicated to sending out The BV Buzz on a regular basis. It's been the one consistent thing that continues to improve our brand recognition that doesn't cost much.

Get on a regular blogging schedule
Just like our email newsletter, we've made it a point to update The Banner Blog on a regular schedule. It's an opportunity to provide something of value to our current and potential customer base, gain recognition from the search engines and position ourselves as experts. Again, it costs us very little and was a small improvement we implemented

in December of 2008. We've received many accolades, it's brought increased traffic to the BannerView.com website and led to multiple speaking engagements, which in turn brought in customers.

Stay visible in your community

Ah, the power of networking. There are so many events going on in a major metropolitan area, that it's just plain smart to get out there and network. However, networking should be done with a purpose. Whether it's trying to land a new client, penetrate the company's brand into the market or simply meet new people, networking can be a low cost, yet invaluable opportunity to get results. If you are reading this blog, there's a great chance you've personally met me at a networking event. Even if you may not be a potential customer, it shows that our brand has penetrated the market, driven traffic to our website and placed us front of mind when it comes to getting your potential referrals for those in need of successfully conducting business on the Internet.

We can continue to utilize and benefit from the above 3 things no matter what the economy is doing. When the timing is right we'll also combine those 3 things with other opportunities that will make for a big impact. However, all along it's those little things that setup the big things to make the greatest splash. In baseball it's the pitcher that gets the credit for a perfect game, but it's the defense making the plays, the catcher calling the right signs and the offense scoring runs so that the pitcher can come out the hero.

Chapter 2
Spending Money in the Right Place Can Get Your Brand on the Radar

You may be thinking "duh" to the title of this chapter, but it's not what it appears.

Yes, spending money promoting your brand via advertising can certainly improve your brand standing, but that's part of your marketing plan. There are other areas and opportunities to place your brand on the radar.

For example, we joined an organization that had a membership fee of $5,000 for the top membership level. We figured that by joining at the highest level, we'd be treated better and get access to more important people than the lower-level members.

Boy, we were wrong. It seemed that we got no better access, respect, or new business from our highest-level membership. It didn't help that the other premier members were some of the largest companies in the area since they could spend additional dollars on sponsorship opportunities and other means to generate respect.

We figured we needed to continue staying active for at least another year. So, year two we invested another $5,000 to keep the premium membership level. Yet again, nothing. No leads were sent our way, we didn't appear to be on the radar of the people who ran the organization, and we certainly weren't being introduced to the top people in the community. There goes ten grand down the drain which, in a small business, hurts like hell!

Like most people, throwing good money after bad wasn't beyond our capability. Since we were "in" $10,000, we said heck, why not throw some more money down the drain. So, we asked about sponsorship opportunities, and one became available for one of the most attended events of the year. The sponsorship was only $1,000. We negotiated a multi-tier sponsorship that included that event and a few other events for a total of $1,500.

Bam! All of a sudden we showed up on the radar of just about everyone in the organization. We noticed an immediate change in attitude towards our company. In fact, we were invited to be presenters at one of their sponsored events. We were even nominated for the organization's annual business recognition program.

Coincidence?

I think not.

Of course we were flabbergasted by all of the sudden recognition. We spent $20,000 over two years and got nothing. Now we were spending a simple $1,500, and we got noticed. Does this organization have some sort of trigger that jump-starts their recognition of companies when certain threshold bells go off? No. It was a matter of spending money in the right place and at the right time.

Membership to the organization, no matter what level, doesn't seem to make a difference. It's simply the cost of admission. Event sponsorship appears to be more important. Different people become more aware of you when money is spent in the their department.

Another example of this in action was a major local media company that runs several publications. In this case, we didn't even have to spend our own money, but garnered recognition and respect from key people within that organization.

I happened to be involved in negotiations for a semi-significant media buy in an issue of one of their publications promoting an organization's big yearly event. I set up the deal to be a win-win situation for both parties involved. The media company, which I had been courting in a number of other areas, all of a sudden had me and, consequently, our company on their radar.

Within only a few weeks of the deal being completed, we were contacted to give a proposal for a special project. While we didn't end up getting the job, people within the organization showed us more respect, and they made sure to go out of their way to introduce us to people in their circle, which helped position us to be recognized for an upcoming special publication.

Here's another example that demonstrates how spending money in the right place can lead to brand success. About 18 months prior to even thinking about writing a book, I saw Jonathan Peters (my editor for this book) at a networking event. He had already been a client, but had simple needs for hosting a website and handling email (translation: he wasn't spending a lot of money).

At the end of that networking event, I said, "let's do lunch."

A few days later, I had my assistant get in touch with him and setup a dinner meeting. His wife and my wife were invited. So the get-together turned from a simple lunch between the two of us into a dinner for four at a very nice (pricey) restaurant. I ended up spending more on that dinner than he had probably spent as a customer over the last few years!

If looking at it from a simple return-on-investment standpoint, I would have been out of my mind. In fact, Jonathan would have to remain a customer for a couple more years just for us to break even on the dinner! My wife and I usually cook dinner (okay, she cooks) and eat at home with the kids. So going out to a fancy dinner wasn't something I'd normally do. For some reason, I had a good feeling about it and had no regrets since everyone had a great time.

While Jonathan continued to remain a client, we hadn't been in touch for nearly a year when we got together at my office. Of course the subject came up of the dinner and he expressed what a nice surprise it was going from a simple business lunch to an elaborate double dinner date. I knew that it made a good impression, and that it reflected nicely on what our company provides. Best of all, Jonathan was finally in a situation to launch a new business venture, and he needed a complete solution. It was a nice surprise for me since I hadn't thought about that dinner until he got back in touch.

Jonathan is spending more money now and the dinner turned out to be a great investment. We've made our money back (got you Jonathan)!

Actually, spending that money on a dinner turned out to be a huge return on investment. Jonathan signed up for a much more complex level of service (translation: he was spending a lot more money). However, that was just the first step to bringing us to where we are today.

A few months after Jonathan embarked on his new Web project, we began a client and prospective client educational and entertainment tour (referred to as the BannerEdutainment Tour) of our company. This was an opportunity for us to educate and introduce our current and potential customers to the services we have to offer and further provide them with a personal connection to our brand.

Jonathan attended one of the events. After that event he mentioned his partnership with someone who was putting together workshops with quality speakers, Jonathan being one of them. I mentioned how we were doing a number of speaking gigs and maybe we could get involved speaking as an expert on leveraging email marketing and blogging to promote a brand online. At the same time, we discussed his area of interest for the seminars, and how he focused on the book publishing aspect that would help speakers, trainers, and coaches build credibility and enhance their earnings potential. Since Jonathan was already a published author and accomplished ghostwriter, I thought to myself, "Hey, maybe that's an avenue for me to build credibility and better position the BannerView.com brand."

A few weeks went by and Jonathan and I coordinated the booking of a flight, handled the hotel reservations, and I prepared myself to join him and the lineup of speakers to put on a weekend-long workshop.

In the meantime, Jonathan and I had a lengthy conversation about book writing. Off I went, typing away on material for this book, which leverages both our areas of expertise.

Looking back at that dinner we had together, would I be writing and would you be reading this book had that dinner not happened? It was money well spent in the right place.

The blog entry below speaks to the fact that your online and offline marketing efforts need to align in order to build a great brand.

The
BANNER
BLOG

Combined Online and Offline Marketing Efforts Lead to Marketing Campaign Success

http://www.bannerview.com/banner-blogs/mark-cenicola/story.bv?storyid=1211

Here's a (not so) surprising study that simply confirms that hype in all areas of life runs rampant, while common sense takes a back seat. Really, eat less calories and you'll lose weight? What a concept!

The weight loss debate about specific diets being better than others is similar to what's going on right now in the online world. The Web, while a powerful medium for marketing, is simply that, just a medium to do these things. The well rounded approach of online and offline efforts with a strong plan of action is still key to marketing success. There are no magic bullets, tricks or special diets to getting low cost (or free) advertising and turning that into huge revenues. Since sites like Twitter, Facebook, LinkedIn, MySpace and Ning are free and get a lot of attention, people assume that online marketing is free and easy.

When it comes to online marketing, these areas are getting a lot of attention:

WSO/SEO

Web Site Optimization/Search Engine Optimization is the process of enhancing a Web site to attract more visitors from search engines and converting that traffic into sales or leads once it hits your Web site. One often overlooked facet of Web Site Optimization is the quality of traffic coming to your Web site and the quality of your Web site in general. It's much better to get targeted traffic to your Web site that is likely to convert to sales and leads then it is to get a bunch of traffic to your Web site and not get any sales or leads. I'm sure you'll take 100 visitors a day/week/month that make a purchase, then 1000 visitors that don't. Also, it doesn't make any sense in spending money to increase traffic to your Web site if your offerings, pricing or ease of purchase once they get to your Web site are not optimized. Focus on the latter items first before you bother with the former. WSO/SEO takes time, anywhere from 3 to 12 months before results can be obtained. Sometimes the competition is so fierce, these efforts may never pay off as there are no guarantees. That's where SEM/PPC comes into play.

SEM/PPC

Search Engine Marketing/Pay Per Click involves paying money every time a person clicks to visit your Web site.

For example, on Google, you have the "Sponsored Search" results on the very top and to the right side of the page (see graphic below). When someone clicks on those links, it costs the advertiser money. Once the money runs out or the advertiser stops paying, those links are removed. The other links, known as "organic links" (those obtained via WSO/SEO) don't cost any money or disappear because an advertiser stops paying. In fact, no amount of money will guarantee that your link will get displayed in the "organic links" section. SEM/PPC works and is popular because it can net immediate results, but in the long run, can be much more expensive and short lived than success obtained via WSO/SEO.

Social Media/Web 2.0
Social Media/Web 2.0 involves things such as Social Networking and Blogging. There are various Web sites from massive social networks like Twitter, Facebook, LinkedIn, MySpace to a popular newcomer called Ning where you can develop your vertical social networks. As mentioned above, these sites are free, but it doesn't mean using them is necessarily free. Whereas having your Web site reconstructed to be optimized and ready to accept business has hard costs as does paying for immediate traffic to your Web site, Social Media/Web 2.0 can have significant soft costs so don't get confused by "free." Some organizations dedicate entire employees just to be active on Social Networking Web sites as that's what it can take to make it successful. I recently visited Zappos.com's headquarters and learned that they have 5 employees alone dedicated to Social Media! You get what you put into Social Media, results

aren't guaranteed and there are situations where it can backfire. The upside is that unlike traditional advertising the benefits are usually long lasting. Brands can be built, goodwill can be obtained and your PR profile can significantly be enhanced. It's very difficult to obtain those type of benefits simply by purchasing advertising. In fact, you can now follow BannerView.com on Twitter.

Educating yourself about the above facets of online marketing can be critical to helping you avoid the hype and expend your efforts in the right area.

Just with any marketing campaign or diet, it takes time, money, patience and a commitment to achieve success.

*****Visit the original blog posting online to see an example of a Google search page and commentary*****

Sometimes it's not about spending money with the goal of directly gaining a benefit. Sometimes it's more about where you spend that money, and who recognizes that money is being spent. Always keep an eye open for where you can spend money to get the biggest recognition as opposed to direct returns.

Of course, spending money in places you need to spend the money anyway, and getting side benefits, is the best of both worlds. Money buys influence. Just make sure that your are influencing the right people.

CHAPTER 3
GREAT BRANDS MAKE GOOD IMPRESSIONS AND ASSUME GOOD REPUTATIONS

*A*s a small business owner, there's not a lot of distinction between your business reputation and your personal reputation. Your brand's reputation usually equals your personal reputation.

Business professionals who already know me, and the brand I represent, think, "Hey that's Mark Cenicola from BannerView.com" when they see me. They associate me with the brand I represent, sometime they even view me and my brand as one in the same.

Similarly, if a person who is familiar with our brand meets an employee from my company, they already have an impression about that employee just by their brand association.

It's like saying you have a degree from an ivy league school; you are assumed to be smart and carry a certain reputation even if you were the worst student that school ever had! The brand's reputation becomes your reputation. You already have the advantage of people assuming you have a good reputation merely by representing a great brand.

On the other hand, people who don't know me, or the brand I represent, have no reason to assume my reputation is good or bad. My reputation is at a neutral state. This makes it harder for me to prove the qualities of having a good reputation regardless of the brand I represent. I prefer having a great brand that precedes me, so I'm assumed to have a good reputation. But when this doesn't happen, I still have the opportunity to make a good impression and build a good reputation.

On the negative side, a bad brand can destroy a reputation. If you represent a bad brand, people will assume your reputation is bad.

How many times have you met someone from a brand you hate? What's your first impression of that person? Can you guess what their reputation is without even getting to know them? It's extremely difficult to overcome the first impression when representing a bad brand. You have to work doubly as hard to make a good impression and build a good reputation. Sometimes it's impossible.

However, a good reputation doesn't always translate into a great brand. It takes multiple ingredients to make a great brand. A great brand is visible, recognizable, and has a good reputation. When people are exposed to a brand that leaves the recipient with a good feeling, that's a great brand. Good feelings are also felt toward those with good reputations.

A small business owner with a good reputation can't be everywhere at once so there's limited exposure opportunities of a good reputation. Brands can continue to work when you can't. They can go places where you can't. They can touch people you can't. Brands can outlive your employment or ownership in the small business. They can even outlast you!

Most important of all, great brands overcome bad impressions and reputations. If you have a great brand that precedes you, the job of being assigned a good reputation is easy. At most, giving a mediocre or neutral impression will allow you to keep your assumed good reputation. You'd have to make a really bad impression to get assigned a negative reputation. And assuming the worst (like making an ass of yourself), people will write you off as an exception to that brand, thereby separating your reputation from the brand's reputation.

That is good news for small business owners. Just knowing that a great reputation can take some hits and that you will be given the benefit of the doubt can be comforting. We've certainly seen our fair share of employees who make mistakes (me included), but part of having a great brand is being able to shrug off those moments without damage to the brand.

If a mistake does happen and you have a great brand, you'll find out about it. Loyal customers will let you know when your brand takes hits. And it will happen. If you think you or your staff has never made or will never make mistakes, you're kidding yourself. All you can do is ensure that your brand is strong enough and your customers loyal enough to weather mistakes and other hits.

Knowing whether you and your brand have a great reputation is half the battle. You can start online by checking out your Google results page. As explained in this next blog entry, you can see that "Googling" yourself can reveal a lot!

The BANNER BLOG

What's Your Google Results Page Look Like?

http://www.bannerview.com/banner-blogs/mark-cenicola/story.bv?storyid=1323

Type your name into Google and see what comes up. Do you dominate the first page of the results? Are the links to pages you control? Are the links what you might want a potential employer to see?

Employers are turning more to the Web and Googling prospective employees before they even call them in response to a resume. In fact, we are currently hiring for a few positions and I've personally Googled a few of the applicants out of curiosity.

Here's what I found when Googleing for Job Applicants:

1) Nothing - many applicants didn't have any Internet trace on the first page of Google when typing in their names. This means they probably haven't posted a public profile on Myspace, Facebook, Linked In or the various other social media sites. As an employer, not having information out there makes me wonder if they've been hiding under a rock or aren't very Internet savvy which is important if you are going to work for a technology company!

2) TMI or Too Much Information - as an employer, I don't care about your wild party escapades last week or your provocative photos in not so flattering ways. A few of the people I've Google had too much personal information that cost them the opportunity to land an interview. If you are out there on the social media websites, be sure to keep it clean and professional because your next employer could see what you post.

3) Too many competitors - if you have a common name it's very difficult to own the first few results in Google and its frustrating as an employer to try to track you down from those results. If you can't be found it's the same as nothing out there about yourself. So, what to do about it? Brand yourself - if you have some sort of personal brand motto, including it in your online profiles or having your very own website for job hunting purposes can make an employer find you easier. And if an employer comes across a professionally done online profile that's easily found, your hire-ability factor just increased over your competition.

The moral of the story is if you are looking for a job, consider your personal brand. Maybe it's time to work on that first and use it as a stepping stone to land that perfect job!

CHAPTER 4
BLOGGING ISN'T FOR DUMMIES

*B*logging is a critical component of branding on a budget. Getting people to follow your story is what branding is all about. If you are successful in this area, it can lead to a number of opportunities to raise your brand's awareness.

What many people may not realize is that blogging is not new. It is probably one of the oldest Web 2.0 technologies because it was really part of Web 1.0. After all, the first web home pages could be considered blogs because they were sites where people told the world about themselves, and that they updated whenever something new happened.

So let's step back a minute and define what a blog really is. Blog, the shortened version of Web log, is simply an on-line diary or account of someone's life. It can also be treated like an opinion article that you might see in a newspaper. The stories in your blog incorporate experiences from your life and your views about a topic.

Here are five elements of a successful blog:

1. **They tell a story** by incorporating real life examples into lessons that can be learned and retold by others. Blogs that simply regurgitate news or facts without some sort of individualized spin don't make for very good reads since there isn't much for people to follow.

2. **They back up stories with facts and cite sources**. The more facts and citations (links to other material), the more credible your stories will become. While you can reference previous blog entries, if you only link and reference your own material people will see your blogs as self serving and quit reading. The Web is built on links to other places. Use them and people will come to you as a resource.

3. **They are focused on a specific area**. A blog that's all over the place doesn't generate a large following. The more you stick to a particular subject, the better chance you have at building loyal readers and followers. It's difficult to be all things to all people, so focus on an area where you have expertise. When you establish yourself as an expert in one particular field, you'd be surprised by how you can leverage that expertise into other areas. Eventually, people outside that area of expertise will latch on and begin to follow your story.

4. **They usually have a theme** and are driven by goals. The purpose of your blog should be established

before you write your first entry. Use that first entry to share your theme, purpose and goals with your potential readers. They'll appreciate your frankness and be able to identify if your blog is something with which they will relate. By sticking to your theme, you will build loyalty with your readers, and they will appreciate the consistent message from entry to entry.

5. **They are complete and intelligently written.** If you are not a good writer, have poor grammar, and can't spell, then by all means, don't embarrass yourself! The line between journalism and blogging is becoming blurred. The reason for that is that it's hard to tell the difference in quality between what great bloggers write and what journalists report. While most bloggers don't have anywhere near the credentials needed to be a journalist, they take care to write blog entries that are well thought out, well organized, incorporate facts, cite sources, are spell checked, and have proper grammar. If you are not a good writer, there's still hope, however. Just make sure to have someone review your entries before you publish them. There's certainly no excuse for mis-spelled words in the day of spell check. A few minor grammatical errors are understandable; after all, I've made them myself. I only know about my typos because my readers are prompt to get all over my case when they find them!

So what are some of the opportunities that blogging can bring? In less than a year from starting my own blog, I had a number of avid readers, people who commented both online and off. Speaking engagements became available because of the credibility my blog provided. I was also creating content for both BannerView.com's email marketing newsletters and, of course, for this book! Blogging is what really initiated my ability to build a great business brand on a budget.

And remember, blogging isn't always about the number of comments you get on your blog. In fact, we get very few comments on our blog. The success of your blog is based upon how it contributes to your credibility, inspires you to do other things, helps you practice what you preach, and many other benefits to your personal brand when done properly.

CHAPTER 5
LEVERAGE THAT BLOG!

*F*inally! This is the chapter where I really get to leverage all the hard work I put into my blog over the last year. Okay, well it wasn't really all that much work. In fact, I spent at most 3 hours a week on my blog activities. Yes, it's not really that much work. There are bloggers who do it for a living and spend quite a bit more time, but my purpose was to use it as part of an overall brand building strategy.

My blog is updated once a week on Wednesday and it's rare that I post anymore than that. This schedule has been consistent since I started the blog in December 2008 and has continued into 2010 without interruption. When I know I'll be unavailable to write a blog, I write entries in advance and publish them on Wednesday which provides flexibility when I'm away and keeps readers engaged because they can anticipate and predict when new content will be published.

Previous to this chapter I have shared with you a few raw entries from my blog. However, in this chapter and in subsequent chapters, you'll see my raw blog entries used more prevalently throughout the content as an example of how it's being leveraged to write this book. The entries will be shared where relevant and appropriate. I've included the URLs so you can visit the original posts, view any comments that may have been posted, or even comment on them yourself.

In the Blogging Isn't for Dummies chapter, I referred to one of the five elements of a successful blog as having a theme and being driven by goals. The following entry was my first foray into blogging. It provides my readers with insight into why I started my blog, the goals, and even has a little rant about Web 2.0 and social networking.

The
BANNER
BLOG

Web 2.0 This, Social Media That, Blog Here, Blog There, Blog Everywhere

http://www.bannerview.com/banner-blogs/mark-cenicola/story.bv?storyid=1156

So it officially begins. Upon some not so subtle nudging from our resident Marketing Coordinator, Brian Mell, I've decided to start my own company blog. I've not been one to keep a personal diary, which is really what a Web log is (blog) just in electronic format, but here's an opportunity to keep a professional diary to share my experiences with customers, potential customers, employees, business partners, business associates and even competitors. After all, this is Web 2.0, so why not?

Before I dive into the reasons behind starting my company blog, let me first debunk the myths of and define Web 2.0.

Web 2.0 is not a new technology and it's not even a new way of using the Web. It's simply the migration toward the reliance on ordinary people as sources for information. As opposed to only trusting information from highly organized entities such as the government, educational institutions or media outlets, people are turning to one another for direct/raw access to information. People of society are now the media, hence the term "Social Media." "Blogs" are the primary features of Web 2.0. However, the nature of the Web has been that blogs were one of the first mainstream features of the commercial Web so it's odd that it's commonly referred to as a Web 2.0 feature. Another subset of Web 2.0, is the migration toward making mass electronic connections to individuals as opposed to more traditional relationship building. This is known at "Social Networking." I feel the term should really be referred to as "Electronic Networking." I'm not a big fan of Electronic Networking as it's commonly misused. Many Web sites such as Facebook, LinkedIn and MySpace exist for the purpose of making social connections. I've refrained from using these services because I do not want people requesting to be part of my social network before I get to know them and I would expect the same consideration. Meeting someone briefly in passing does not qualify someone to be part of my social network nor is it necessary to publish all of my close relationships for me to be used to pillage them. Since not all relationships are created equal and there are only so many close relationships a person can possibly maintain, I simply decline all invitations to be part of an electronic networking Web site. Therefore, don't take it personally if you send me an invitation to be in your electronic network and I decline. Exchanging regular dialog with me, whether it be via comments on my blog, via email, by telephone and/or in person are all ways to build a relationship and be included in my social, yet still private, network.

Now that I have my Web 2.0 rant out of the way, I'll share the **reasons behind this blog:**

1) It's a way to forge new relationships with other professionals and get feedback.

2) It's a forum for others to share my expertise in conducting business online.

3) It's an opportunity to keep my writing skills fresh.

4) It's an opportunity to build a following, albeit somewhat self promoting.

5) The fresh content certainly helps with Search Engine Optimization for BannerView.com.

The **primary goal of this blog** is to generate new business opportunities for BannerView.com.

What can you expect to find in my blog?

1) You'll find articles that are relevant to conducting business online.

2) I'll share my personal experiences from various industry events which have relevance to conducting business online.

This will include photos and sometimes video depending on how ambitious I feel.

3) You may see me feature our clients from time to time. I hope to include their success stories which may benefit you in conducting business online.

4) You'll hear some of the stories from behind the scenes.

5) You'll see other information which may be relevant to my role in helping companies and organizations of all sizes build, promote and maintain their online presence.

As the President & CEO of a company that helps other companies do business on the Internet, you may be asking what took me so long to start a blog. The easy answer is time. Maintaining a blog is a commitment. It takes time to write quality content that people will want to read and it takes a commitment to writing new things on a regular basis. I have to give credit to my fantastic staff, especially to my assistant, Lisa Schonsheck, who has taken on many responsibilities that have freed me to do things such as this blog! Sign up for the RSS feed and check back often for new postings!

Sincerely,

Mark Cenicola

This next entry was inspired by my instant rock star status achieved after my first blog posting. Well, not exactly rock star status, but it was nice to know people were reading!

The
**BANNER
BLOG**

If you Blog It, They Will Come: Success in Professional Blogging

http://www.bannerview.com/banner-blogs/mark-cenicola/story.bv?storyid=1168

Whether writing a book, newspaper column, advertising material, email, blog entry or any other written piece of work, the question is: What's the point in writing it if nobody is going to read it? That was one of my main concerns when I decided to start my professional blog. I had to assume the worst and that it was not going to be read. If that was the case, a primary reason, as stated in my initial blog entry, would not be met creating a major deterrent to continuing the effort.

Within hours of my first official blog entry, I already had my first comment which was very encouraging. Thanks goes out to "ipodflea" for participating in a constructive debate. The following day after my first entry, I was at In Business Las Vegas' 50 Most Influential (links to a PDF) presentation ceremony and cocktail party. A gentleman approached me and said he had read my first blog post and complimented me on my entry and writing abilities. I was flattered and quite surprised. While I still have a long way to go before my blog can be claimed a great success, as this is only two confirmed third party blog readers, it goes to show that having a professional blog can attract attention which makes the effort all the more worth it.

In this blog entry, I'm going to cover some of the things that should contribute to the success of a **professional blog**:

1) Write Quality Content - On the Web, content is king. While not everyone is an English major and entries won't be perfect (mine included), checking for basic spelling and grammatical errors and proofreading your posts should be a given. Having quality content in your blog should keep readers interested in what you have to say.

2) Provide a Value Proposition - Tips, advice and/or links to other resources, from which readers can benefit, will provide them with a reason to come back. This is also a great way to get readers to recommend your blog to others.

3) Manage Reader Expectations - Clearly define the purpose of your blog early. This includes using a relevant entry title or subject heading. Readers may come to rely on your blog as a source for specific information. Let them know what to expect and deliver on those expectations.

4) Cutout the Fluff - Most readers won't care what you had for breakfast, the song that was playing on the radio while driving to work or your gripes about the weather. While readers want to be entertained, they are busy and will appreciate you getting to the point quickly. Although, if you're Paris Hilton people might care about what you had for breakfast, but remember that I'm referring to a "**professional blog**," not a personal blog.

5) Read & Respond to Comments - If your readers are going to take the time to comment on your blog, the least you can do is respond back to them in a timely fashion. Why should your readers spend the time to comment on your blog if you aren't spending the time to read and respond to their posts? While you may not be able to read or respond to every single comment, try to drop in every once in a while to let your readers know that your care.

6) Be Prepared for the Unexpected - Starting your own blog and allowing comments on your blog entries can bring unexpected responses. Since not everyone is going to agree with everything you have to say, be ready for harsh criticisms. However, this can be an opportunity to engage constructive debate which will increase blog activity. On the other hand, be ready to handle praise with humility.

7) Monitor For Unwanted Posts - It's important to regularly monitor your blog for unwanted posts, removing garbage from getting in the way of real discussion. If you have a popular blog, it may become a spammer's playground as they try to redirect your reader traffic. Unscrupulous readers may use your blog to comment with indecent material, offensive language or personal attacks. These types of posts should be removed quickly.

Having a blog is one thing, putting in the effort to make it successful is another. Hopefully, the above advice will put you on the right path to benefitting your online business by utilizing the power of blogging.

This next entry relates how blogging should be viewed from a PR and branding perspective as opposed to a marketing and advertising perspective.

The BANNER BLOG

Professional Blogging to Generate Business: The Power of PR & Branding

http://www.bannerview.com/banner-blogs/mark-cenicola/story.bv?storyid=1174

Right after the latest edition of The BV Beat was delivered to subscribers' email boxes, a customer emailed to ask if they should start a blog to improve their Web site. Being in the business of building, promoting and maintaining a company's online presence, my gut reaction was to reply, "Of course!" as that would have translated into a revenue opportunity for BannerView.com. Knowing that advising our customers on the best long term strategy for their e-business operations has consistently paid more dividends for both them and our company, I wasn't so quick to respond with a definitive "yes" and diverted the conversation toward gathering more information.

Obviously, I've been warming up to idea of blogging and there continues to be a proliferation of both personal and professional blogs. Therefore, a "yes" answer to the customer's question would certainly seem reasonable. Wait! Not so fast. Even though receipt of positive feedback has certainly helped my warming process to blogging, I can't yet say it has generated direct business for the company. It's only been a few weeks since I launched my blog so time will tell, but my expectation is that this blog will not result in direct sales for the company. That statement may seem hypocritical of my first blog entry as I mentioned that the primary goal of this blog is to generate new business opportunities for BannerView.com. That goal still holds true, but my expectation is that those business opportunities will be more indirectly achieved, not from me hawking our company's wares directly from this blog.

Blogging should be viewed as a form of **public relations** (PR) and **branding**, not **marketing** and **advertising**. Understanding the differences between those types of promotion is key to understanding how to use blogging to generate business for your company. I found the following illustration posted in a blog entry on Ads of the World that serves as a great example of explaining the differences between PR, branding, marketing & advertising:

As you can see from the illustration on the following page, PR and branding are indirect forms of getting people to buy your products and services, whereas marketing & advertising are more direct forms. So why should blogging be viewed as PR & branding? Since you probably already have an entire Web site devoted to telling people why you are the greatest company in the world (marketing & advertising), why clutter your blog with more of the same propaganda? Your blog is a great way to position yourself as an expert in a particular field by providing unique and valuable content to which others can refer. If you're simply regurgitating the same self-promotion that's contained on

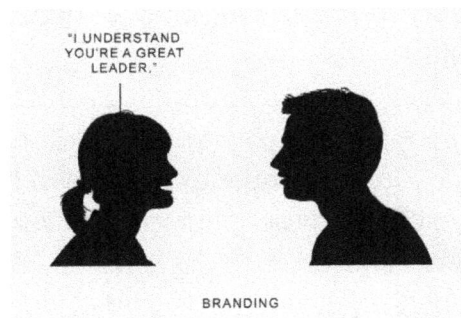

63

your Web site you'll have a hard time getting people to follow your blog.

For the best chance of success, the various methods of promoting your products and services need to be examined prior to embarking on a promotional campaign. Many companies immediately look to marketing and advertising as it appears to be easier and faster to implement, but the costs are usually high and the effects short lived. PR and branding can take much longer and be more difficult to implement, but the effects are lasting and it increases the likelihood of success of your marketing and advertising efforts.

So the answer to the customer's question of whether or not they should start a blog to improve their Web site really raises additional questions. Are you going to take the time to implement a PR and branding campaign of which a blog can be a valuable tool or are you more interested in a quick promotional hit via marketing & advertising for which a different tool may be better equipped? This blog is a leg of a bigger PR and branding strategy and for it to have a chance at success, the proper time and effort must be given to it. If you are ready to devote the resources necessary to make your blog a success, then yes, let's get you started!

After a couple of months of blogging, I wanted to do a test to see if people were still reading the blog, so we ran a little contest.

Blogging for T-Shirts

http://www.bannerview.com/banner-blogs/mark-cenicola/story.bv?storyid=1189

It's been more than two months since starting my professional blog and making my first blog entry. Have I seen any benefits? My initial entry received a number of comments, but subsequent posts haven't been as active. I've had people approach me to compliment some of the articles written and received emails from people that have liked particular entries. It appears that people are reading even though they might not be commenting. The takeaway is that my personal brand awareness has increased from these blogging efforts given me comfort that there continues to be potential value in keeping my commitment to at least once a week updates.

It would be nice to see comments to my blog entries, in fact all blog entries at BannerView.com, increase. That would certainly help get some of the others in the organization committed to starting their blogs and updating them regularly. I'm asking for your help to increase comments and if this works, will serve as a tip for getting people involved in your own blog, albeit a slight gimmicky. As an incentive for asking for your help, I'm going to present you with a small gift of appreciation for participating. How does the official BannerView.com T-Shirt sound? That's right, comment on our blogs and I'll send you the Official BannerView.com

T-Shirt! Here's your chance to get this gorgeous black t-shirt, which matches just about anything, for FREE before it goes on sale to the general public.

To claim your Official BannerView.com T-Shirt, you must post a comment anywhere on the blogs between February 11, 2009 and February 25, 2009. It doesn't have to be a comment on this particular entry. You can post comments on any blog article. Don't post your email or physical mailing address in any of your comments requesting your t-shirt. Instead, after you've left a comment, use the form on the contact page to tell us that you've posted a comment on the blogs, you want your Official BannerView.com T-Shirt and provide your full shipping information. Shirts are available in sizes small, medium, large & x-large so make sure to include what size you'd like. Once we verify that you've posted a comment (provide us with a link for quick verification), we'll send you your t-shirt. Not to worry, you won't get hit up with sales calls or other garbage, just the t-shirt. Of course, we have a limited number of BannerView.com T-Shirts available so first come first served and only one per person that posts a comment. When we run out, that's it! You'll be stuck waiting until they go on sale! Even if you don't want the Official BannerView.com T-Shirt (hard to believe who wouldn't), your comments are still welcome! Free registration is required before commenting. If you haven't already, you can register here.

For those of you that are curious, I'll report back in a few weeks to let you know if my little experiment worked in increasing activity on the blogs.

*****Visit the original blog posting online to see the photos*****

This next entry was more of an announcement of success we had from our blogging efforts.

The BANNER BLOG

Blogging Overload is Delivering Website Traffic

http://www.bannerview.com/banner-blogs/mark-cenicola/story.bv?storyid=1248

I've been blogging, speaking about blogging and now writing in my blog about blogging! It seems that I've hit blogging overload, but in a good way. I've promised to periodically update you on how this blog has been performing since its official launch in December 2008. The effort seemed to hit a lull in March and April, but things came roaring back in May and June.

The previous success of this blog has inspired Jeff Helvin to jump on the blogging bandwagon with his "Blogging for Balance" view. In fact, after a couple of months it seems he's hit his stride. His May 27th entry, "What Do You Have Waiting in the Wings?" hit a high note with multiple clients and friends of BannerView.com prompting direct feedback related to that specific entry.

The latest set of blog articles that were featured in the June 3rd edition of The BV Buzz generated multiple lead inquiries over previous editions and our lead generation activity continues to improve due to the efforts of the blog. As mentioned, one of the primary reasons for professional blogging is to help drive traffic to the BannerView.com website with the intent of turning that traffic into leads and ultimately new business for the company. Therefore, I'm quite pleased in the overall success of our blogging initiatives and am highly recommending our clients do the same.

The biggest mistake we see companies make is giving in to quickly or expecting results too soon with their websites. Even when you hit a lull with your efforts, which is bound to happen from time to time (anybody watching the economy?), persistence can pay off. Stay focused at the goals at hand by observing what happens during the downtime, learn from it and use it to continually improve your strategy going forward.

A big challenge of committing to a consistent blogging schedule, is coming up with ideas to get the creative writing juices flowing. The next entry provides some tips on generate ideas, and it includes a little self promotion as a lead to new blogging ideas.

The BANNER BLOG

Tips to Finding Things to Write About for Your Blog I The BV Ride A Part of History?

http://www.bannerview.com/banner-blogs/mark-cenicola/story.bv?storyid=1280

One of the hardest parts of keeping to a consistent blogging schedule is finding things to write about as constantly coming up with new ideas can be challenging. If you've ever been thinking of starting a blog, keep in mind that consistent updates is one of the keys to success. No one wants to read a blog that never gets updated nor if updates are sporadic and unpredictable at best. So how do you go about finding new things to write on a regular basis?

It's easy when you have stories to tell. Everyday life is a story, but finding the right angle on the things you do on a daily basis is the key to creating unique stories worth telling. People love a good story, maybe one that has a lesson to be learned or simply something interesting. The really good stories are often retold to others, thereby increasing the chance that someone will pass along a link to your blog entry which results to an increase in website traffic. While theory is important, it's the life stories that the majority of people find interesting. If you never leave your home, office or step outside of your normal day to day routine, you probably won't be a very interesting blogger.

Blogging has forced me to do interesting things outside of the everyday norm so that I would have interesting stories to tell to create a better reader experience. It's also provided the opportunity to direct people toward the blog with photos and additional commentary to backup a story I've told. If you find yourself in a rut with nothing new to write about for your blog, do something new and especially out of the ordinary. The BV Ride is just one example of something out of the ordinary that has provided plenty of mileage (no pun intended) for my blog.

The BV Ride a Part of History?

Two weeks ago, the family and I went on a road trip through 6 different states and drove the new BV Ride more than 2,000 miles over that trek. It was a great experience where the journey was just as fun as the destination. We certainly got a lot of looks due to the BV Ride and overall it was good for the BannerView.com brand as much as it was fun for the family.

One of the destinations on our trip was Yellowstone National Park. In fact, during the month of July, 2009, Yellowstone reported a record number of visitors. It was great to see the huge number of visitors, see nearby hotels sold out and see the neighboring tourist towns bustling with activity. It's been quite some time since I've visited any national parks outside of Nevada and the activity was much greater than I remember or had anticipated. It sure does seem to lend credibility to calling the end of the economic down cycle.

Catch the BV Ride in Phoenix as Jeff Helvin and I make a trip to co-present *Effective Use of Blogging & Email Marketing to Drive Website Traffic* brought to you by BusinessWire.

*****Visit the original blog posting online to see the photos*****

So maybe you are liking what I write in my blog. If so, go ahead and capture it's RSS feed. My blog can be found easily within the other BannerView.com blogs by going to www.thebannerblog.com. Even if you don't like my blog, there are other bloggers from the same company who have different topics on which they blog.

What I really want you to take away from these examples is the idea that you can leverage what you do with your blog to build a great brand!

CHAPTER 6
EMAIL MARKETING: BRANDING FRIEND OR FOE?

*E*mail marketing, when done properly, is one of the single most powerful branding tools that can be utilized. The Direct Marketing Association predicts that for every $1 spent more than $39 will be returned on investment well into the 2010's even with the expected increase in spam and other garbage that reaches inboxes.

According to a 2009 survey by Epsilon, an email marketing company, "More than half (57%) of American consumers have more positive opinions about companies that send them emails, and 50% say getting email increases the likelihood they will purchase, either online or offline, from these companies."

What stands out the most about the results from this survey if the fact that customers will purchase from a company offline because they received an email online. This certainly points to the fact that branding, whether online or off, needs to be an integrated strategy.

Furthermore, according to eMarketer, "Email was the top channel for distributing content to friends, with 46.4% of all shares. Content shared by email was most likely to lead to purchases, subscriptions and other conversions." What this means is that emails, being easy to forward, are still one of the most effective ways to spread your brand's message.

It Keeps your Brand in Front of Your Customers

Well written, informative email newsletters are a great way to offer value to the recipient. You must make sure you are providing something of value, not to yourself as the sender, but to the recipient. Usually this is free information that's relevant to the industry to which you serve.

Recipients, while they aren't going to read every email you send, will maintain their subscription to your newsletters if they contain something of value. This also helps keep recipients engaged with your brand and ultimately leads to them sharing your brand's message with their contacts.

In order to successfully keep your brand in front of customers and build brand loyalty with email marketing, all of your efforts must maintain CONSISTENCY.

In the rush to take advantage of email marketing, small businesses make the mistake of choosing any 'ol email marketing service. They sign up, pick a template, import their email list, and start emailing. What this does is send an inconsistent brand message. Picking a boilerplate template that's not specific to your brand is a huge mistake.

In addition, many of these third party services don't fully integrate into your own website. When people subscribe (or even unsubscribe), your branding image should maintain CONSISTENCY. Therefore, it's important to choose an email marketing method that fully integrates with your own website, and get a custom designed email marketing template that properly represents your brand.

Now, it's okay to brand your email newsletter with it's own name. In fact the BannerView.com email newsletter is actually called The BV Buzz. When we send our email messages, The BV Buzz is always in the subject line so that recipients know what to expect. But the most important aspect is that recipients can immediately tell that The BV Buzz is connected with the BannerView.com brand, which essentially reinforces the BannerView.com brand.

You also need to maintain a CONSISTENT email marketing schedule. Part of keeping your brand in front of customers is regularly putting that brand in from of them. It's critically important to not only set a schedule, but also to maintain that schedule at all costs. When you send out your email marketing newsletter at random times or wait many months in between, people will forget about your brand. Instead, they'll choose a competitor's brand that is CONSISTENTLY in their inbox.

There is a company I have done business with and because of that, ended up on their email list. The first few newsletters I received had relevant content that was useful to my company. I enjoyed reading the articles and looked forward to the next emails because of the valuable business ideas that incorporated their services. It also kept their company front of mind and I actually looked for new ideas to use their services. Overall, their emails were well prepared, had consistent branding throughout, and offered valuable content—all the ingredients for success.

However, after several months, I couldn't figure out a pattern or method to their madness. They would go weeks without sending a single email, then hit me a couple of times in the same week. The quality of the content also started to deteriorate in the subsequent newsletters. While it's not always easy to come up with great content for every newsletter, I had come to expect the same value found in their early newsletters.

While I haven't unsubscribed from their newsletter, their brand has taken a back seat in my mind. Yes, I'll research their services when looking for what they provide, but they are no longer doing an effective job of getting me to think of creative ways to spend money with them.

That's why CONSISTENCY is key to a successful email marketing campaign. It's also critical that something of value continues to be the main focus of your newsletters. Not surprisingly the company's services started out great, and then deteriorated along with their email marketing efforts. Could a company's consistency in branding predict the quality of work they provide? You can be the judge of that one.

Promote Sales and Events

With a consistently scheduled email marketing newsletter, the ability to send emails promoting specific sales and events outside of the regularly scheduled newsletters is a great attention grabber. Recipients will recognize the brand that's sending them those emails, but they will also see that offer emails are somewhat different from the regular newsletters. This provides additional attention to sales and events.

It's also important to note that promotional sales and events emails should look slightly different from your regular email newsletters. However, it's still important that brand consistency is maintained throughout your email marketing efforts, so that recipients can immediately identify the special offer with your brand.

Obviously, you want customers to buy your products and services as well as attend your events. Using email marketing for this purpose is critical, but it is only effective if done as a secondary purpose to email marketing.

The frequency at which emails promoting sales and events are sent should be contemplated carefully. Bombarding people's inboxes with too many emails will surely cause them to unsubscribe. If they don't unsubscribe, the impact of the message will be reduced.

The biggest mistake people make with email marketing is not using it as part of the overall branding strategy. More often, it's used only to promote sales and events that benefit only the sender instead of focusing on bringing value to the recipient.

That's just the case with the company that was sending me emails. Not only did subsequent emails have poor quality content, they began to send more promotional emails. It smelled of desperation and lowered my view of a brand I really liked. While one of the benefits of email marketing is promoting sales and events, it's best to approach email marketing as a branding tool which will give those promotional sales and events emails greater impact on the recipient.

So go ahead and utilize email marketing for promoting sales and events, but remember that you will get the most success from using it mostly to send an email newsletter that contains valuable, quality content. Failure to put great content first will cause your promotional emails to get ignored and go straight into the junk or trash folder. Worse yet, subscribers will unsubscribe!

Drive Traffic to Your Website

The idea behind your newsletter is to get people out of their inboxes and to your website. How you do that is by teasing recipients with the content of your email newsletters. I'm not referring to porn, but who knows, that may be helpful to a certain audience. All kidding aside, what I mean by teasing them is to give them a taste of the content to peek their interest, and make them click to read more.

For example, The BV Buzz is made up of the latest blog entries written within our company including mine. The only thing we include is the title of the blog entry, a summary and a link to read more. This requires the visitor to click on the link, leave their inbox, and end up on our website. From there, we now have the reader's attention and have a better chance of getting that person to click more links within our website. Having them on our website provides us additional opportunities to get that person as a customer.

It took us many years to realize this. We used to send out a newsletter with a single topic fully contained in the email. We figured that by incorporating the template of the website, readers would be inclined to click through to our website. Well, that was never the case. They read the article, got the information they needed and moved on to the next email.

If you provide everything that they need in the email like we used to do, there's no reason for customers to visit your website or come into your store. With the amount of emails pouring into a person's inbox, giving them everything they need to be satisfied means they can click delete and move on to the next email. Finding creative ways to get them out of their inbox is a key to successfully driving traffic to your website and engaging them with your brand.

Driving traffic to your website will not be possible if you can't get your emails to arrive in the reader's inbox in the first place. This is one often overlooked part of the equation. If the email gets filtered as spam before arriving or winds up in the junk folder, it's not going to be read and your brand will not get in front of potential customers.

The key to getting emails into inboxes is being a trusted source. One problem with using a third-party email service is that they are not considered trusted sources of email for your company.

For example, if your website domain address is, say, QualityNewsletterContent.com, and the email from which you send your newsletters is newsletter@QualityNewsletterContent.com, why would someone trust your email coming from 89jkjdi28393@somethirdpartyvendor.com? Well, they wouldn't. It's also how a lot of email servers determine whether or not to accept that message in the first place. Therefore, it's critical that your email newsletters come from servers that are authorized to send emails for your domain name.

Not to pinpoint any one particular email provider, but one of the largest such providers had their emails automatically deleted by a major employer. This meant that any company using that provider would never have their email newsletters delivered to the inboxes at that employer.

Getting others to trust the source of your email can be done by having an email marketing program that integrates directly with your web server. In addition to the branding consistency benefits previously mentioned, your emails have a better chance of showing up in the reader's inbox.

It's also important to not go cheap on your website hosting provider. Many low cost web hosting providers end up with customers that don't follow proper email etiquette and can get themselves blacklisted. Once your email server is blacklisted, any emails you send will be rejected. Going with a provider that offers business class web hosting services or getting a dedicated server of your own can help prevent your appearance on blacklists.

One of our customers requested we cover the ethics, laws, and tips to success when it comes to email marketing. Therefore, I blogged about this very topic as you'll see in the following entry. I also included references to source materials, especially since I'm no lawyer, and laws were discussed.

The
BANNER
BLOG

The Ethics, Laws and Tips to Success with Email Marketing

http://www.bannerview.com/banner-blogs/mark-cenicola/story.bv?storyid=1191

A customer recently requested that an article about email marketing, the ethics behind it, what to watch out for, etc. be written. Email marketing is a topic we've covered frequently over the years so now is a great opportunity to revisit that topic to make it relevant for today (and another chance to prove that we usually listen to what our customers want)! So here you go, an article about email marketing, the ethics behind it and what to watch out for, etc.

Enhanced brand awareness, improved company credibility and increased sales are a few things that can be accomplished when using email marketing properly. When not used properly, just the opposite can occur such as inconsistent brand identity, damaged company credibility, reduced sales and even legal problems. To obtain the benefits of email marketing and avoid the pitfalls, here are 8 tips to email marketing success.

1) Brand Consistency
When it comes to branding, consistency is the key to success. When it comes to email marketing, your messages need to be consistent with your brand. How many times have you received an email marketing message that doesn't have the same look, feel and consistency with the brand from which you received the message? It's often overlooked as businesses will rush into email marketing without understanding that this is just another way to deliver your brand message. While your email messages don't have to look exactly like your Web site, they should at least include your logo, slogan and be styled appropriately for a good fit with your existing online identity. This will ensure

you send a consistent brand message to your potential customers helping build your brand awareness. Whether you decide to have your email marketing layout designed in house or outsourced, spend the extra money and effort up front to get something customized to be consistent with your existing brand.

2) Substance
Monthly newsletters are an outstanding venue for letting current and future customers know what's going on within your company and informing them of upcoming events, promotions, and discounts. Its a way to stay in touch. But you must keep in mind the monotony of a hard sale and blatant pitches. It is crucial to keep the customer's best interest top-of-mind. Provide useful industry information, new topics or reports from which they can benefit. If you think of the customer's needs and provide them with content of substance, your efforts won't go unnoticed. Readers will stay subscribed and refer your newsletter to others.

3) Spelling & Grammar
By all means, proof your newsletters and do basic spell checking. While sometimes errors do slip by (our readers are eager to let us know), a considerable number of errors can display a lack of professionalism harming the company's credibility.

4) Predictability
It doesn't matter if you send a weekly, monthly or quarterly email newsletter, having a predictable schedule will help readers anticipate communication from you. People

are less likely to unsubscribe from your lists when they come to expect communication from you on a particular day of the week or month. While your readers may not read every newsletter, they know they can catch the next one as it will come right on schedule. If your schedule is unpredictable so too will be your results. Special promotions or announcements outside of your regular schedule can be used to grab additional attention.

5) Opt-In

This is one area where businesses are easily confused and that can lead to bigger problems. Building an email list is not done overnight. It takes time to get people to subscribe to your list. While you may want to go out and a buy a so called "Opt-In" list to boost your subscriber base, that's a big mistake. Those email lists are often harvested from Internet sites using an automated program. You've probably had this happen to you if your email address ever appeared on a Web site, in a chat room, in an online membership directory, a message board or newsgroup posting. Even if someone did "opt-In" to a list at some point, they didn't opt-in specially to your list if you had to buy it. Using a purchased list can get you labeled as a spammer. Legitimate email marketing providers will not allow the use of purchased lists (BannerView.com included) for use in your email marketing campaigns. While you can comply with Federal Laws, specifically CAN-SPAM (point #6) by purchasing a list, you can be deemed a spammer if someone complains to your provider. Getting shutdown by your provider can be a killer to credibility, especially if someone tries to visit your Web site while it's offline!

6) CAN-SPAM

CAN-SPAM is one of those laws that confuses many. The reason is that most people think it was made to can (as in put a lid on) spam. Instead, what it really means is that you are free to spam as long as you follow the rules of the law. That may seem counter intuitive to pass a law that says, yes, you can spam people, but essentially that's what it says.

Here's a rundown of the law's main provisions:
It bans false or misleading header information.
It prohibits deceptive subject lines.
It requires that your email give recipients an opt-out method.
It requires that commercial email be identified as an advertisement and include the sender's valid physical postal address.
In theory, you could buy a list, harvest email addresses, steal them from others or do whatever you need to do in order to collect email address as long as you let people unsubscribe, don't deceive the recipient and provide your valid postal address to spam them under the law. The law isn't useless by any means as most people who spam are trying to deceive, can't provide legitimate physical mailing addresses and don't honor unsubscribe requests. There have been prosecutions already under the law. However,

simply complying with CAN-SPAM laws can't protect you from having your Web site shutdown by your provider and not hurt your company's credibility by being labeled as a spammer (point #5).

7) List Building

So how do you go about building your list if you can't buy one and still comply with federal law? Provide an easy way for people to subscribe directly from your Web site, provide great content so people will refer your email marketing messages to others and exchange business cards via networking. When networking, always make sure to ask for a business card when you meet someone. Most people have their email address on their card. If they give you their business card, it's an invitation to contact them as long as you comply with CAN-SPAM. I've never had anyone complain about me adding their email address to our list as they can easily unsubscribe, know the message is of a commercial nature and have access to our legitimate physical mailing address. If you happen to exchange business cards with someone more than once or have multiple representatives within your company that network, be sure not to add the same email address twice or add the person again if they've previously unsubscribed. Our systems are automated so when we import into our list, it knows those that are duplicate and flags previous email addresses that have been previously unsubscribed.

With any successful email marketing campaign, the goal is to keep existing subscribers and grow your subscriber base, not tick off people by sending them garbage. Properly planning your email marketing strategy, selecting the right partner for the technology and creating something worth reading are the keys to success.

8) Third-Party Email Marketing

Third-party email marketing can be a great way to utilize email marketing to reach an audience in which you don't already have a direct relationship. Here's how it works. A company has a list of people subscribed to their newsletter that fits your target audience. They offer to sell you ad space in their newsletter or send an entire email to their list on your behalf promoting or endorsing your products, services or even the existence of your own email list so they can subscribe directly. Just tread very carefully on how this arrangement is structured making sure that this third-party company is sending out an email to a list where they already have an existing relationship with their subscribers. The email needs to come from that third-party company directly, including the from address, unsubscribe information and full contact information. The last thing you want to do is hire a company to spam for you because that can cause a world of trouble. Companies (specifically editorial publications) do offer advertisers or partners to sponsor an email to their database when the content will be relevant to their audience. Their database is never seen by the advertiser so they still maintain complete control (and responsibility) for their list.

If the customer (you know who you are) that requested this article found value in it, others probably will as well. Comment below to take credit for your inspirational request and you can be the newest owner of the official BannerView.com T-Shirt!

Resources Utilized in Writing this Article:
Federal Trade Commission Web Site
http://www.ftc.gov/bcp/edu/pubs/business/ecommerce/bus61.shtm

March 6, 2002 BannerView.com Newsletter - Email Newsletter Marketing
http://www.bannerview.com/newsletter/archive/?id=26

January 22, 2003 BannerView.com Newsletter - Company Credibility - How to Achieve it on the Web
http://www.bannerview.com/newsletter/archive/?id=60

April 28, 2004 BannerView.com Newsletter - What Can Be Done About Junk Email?
http://www.bannerview.com/newsletter/archive/?id=101

First Prosecution Under CAN-SPAM Act Returns Guilty Verdict
http://www.govtech.com/gt/125749

This next blog entry, while not specific to email marketing, shows the effectiveness of combining multiple branding methods to help make closing the deal easier.

Hitting Them from Every Angle Makes Closing the Deal Easier

http://www.bannerview.com/banner-blogs/mark-cenicola/story.bv?storyid=1286

While sitting in the company break room having lunch yesterday, our Senior E-Business Advisor, Paul Helvin, walked by to explain how he was meeting with a potential client that attended a recent presentation where I was one of the featured speakers. Of course, that's always nice to hear as one of our strategies of late has been to get ourselves in front of audiences, arm them with the information needed to be successful online and obtain a client or two. Education is a key component to helping companies conduct business online, so when potential customers come to us better informed, it makes our jobs easier.

I told Paul, that in fact, a new client just signed up last Friday that attended a presentation earlier in the year where I was the featured speaker. However, it wasn't because of the presentation that we landed the deal. The client was actually referred to us by a colleague. Without that added referral pointing in our direction, the presentation didn't hold much weight on its own. The client did recall my presentation and had also been a recipient of The BV Buzz, BannerView.com's bi-weekly email newsletter, so that trifecta made closing the sale pretty easy.

Paul said it seems that even though we are doing a lot of good things online right now, it's the word of mouth referrals that are paying the biggest dividends. While that may always be true in most businesses, without the presentation and email communication, the deal wouldn't have been closed as easily or possibly at all! That's why it's important to make sure that you hit your potential customers from every angle including online and off. There have been many cases where a customer has learned about us offline, visited our website or received an email and decided to do business with us. The opposite has been true as well where someone came across us online and it wasn't until they saw an advertisement offline that they decided to call.

Building trust with potential customers takes time. They want to know that you're real, that others trust you and that you provide the information they need to do business with you. Seeing you in person or your advertisements offline shows them that you're real. Having others that recommend you or people they know that do business with you builds trust. Having an easy to use and understandable website provides them with the information to do business with you.

The moral of the conversation was that it's important to be everywhere your potential customers may be as you'll be much more likely to land those customers. Go back and review your online strategy. If you aren't getting new leads from your website, you might be better served using it as a tool to help you close the deal. Or, if you are getting a lot of inquiries because of your website, but no sales, using different offline options may be what you need to close the deal.

Related Blog Entry: Combined Online and Offline Marketing Efforts Lead to Marketing Campaign Success

---

This next entry is one of my favorites, and I incorporated it into my own live presentations on Building Your Personal Brand Using Blogging and Email Marketing. That presentation inspired a lot of content for this book.

The
BANNER
BLOG

Don't Let Your Email Marketing Suck

http://www.bannerview.com/banner-blogs/mark-cenicola/story.bv?storyid=1337

According to the Direct Marketing Association's (DMA) 2009 report, email marketing returned more than $43 for every $1 spent on advertising in 2009. The DMA goes on to predict that email's return on investment will be over $42 for every $1 spent in 2010. As more people jump on the email marketing bandwagon, the ROI will drop slightly. However, as you can see it's still going to be a much more profitable way to spend your advertising dollars than most other mediums. The problem is that the email you are sending probably doesn't get the best return it possibly can.

Here are 5 reasons why your email marketing might suck:

1) You don't utilize email marketing - This may seem straight forward, but there's a good chance you aren't taking advantage of one of the most profitable forms of advertising. It's time to start building your email marketing list and planning for your own email marketing efforts now. Don't be left at the starting gate while your competitors continue to take advantage of email marketing.

2) You send out garbage - Is what you are sending really what people want to read? A good email marketing campaign should deliver value, not just advertisements. You want your subscribers to stay subscribers so give them something for nothing! That something will keep them interested and open to your sales pitches.

3) You give too much - Does your email marketing piece include the full details, article or offer right in the email? If so, you aren't giving your subscribers any reason to do anything further. The goal is to get them out of their email inbox and into your store, website or to pick up the phone. What you can do is tease them with article headlines and summaries instead of full articles or provide partial offer details with full details available at your website. If you give them reason to click outside of their email inbox and onto your website, you've provided yourself another opportunity to engage the subscriber and get them to ultimately purchase your products. You can't do that if they are stuck in their inbox!

4) Your brand isn't consistent - If you've invested the time, money and effort into email marketing, the least you should do is make sure your email marketing pieces are consistent with your branding. All too often I get email marketing pieces that have a generic looking template that doesn't deliver the brand's message properly. In many cases, I don't recognize the email which certainly doesn't provide much brand recognition. Email marketing is one of the best and most cost effective branding opportunities you can use. However, if your email marketing is disconnected from your brand, stop now and rethink your strategy.

5) Your emails don't get to subscriber's inboxes - If subscribers don't get your emails or if they are delivered to the junk folder, your ROI is certainly going to be negatively impacted. Can they really be called "subscribers" if they don't get their subscriptions? The problem is that many third party email marketing services use a bulk set of servers that can often get flagged as spam or if you are on a shared hosting server where someone has been flagged for spamming, your emails too will get flagged for spam. The best way to get your emails delivered is to have the emails sent from the same servers that are authorized to send email for your domain. This requires an integrated email marketing product that is hosted on the same servers as your regular website and email. In addition, make sure you are using a "from" email address associated with your domain. This builds credibility in the eyes of the subscriber and can help get your email delivered into the inbox!

Your email marketing doesn't have to suck and with a few tweaks and suggestions, you can take advantage of the low cost email marketing opportunities that can deliver a big return on investment, help build brand awareness and engage potential customers.

If you are interested in other related email marketing statistics from the 2009 DMA report and don't want to shell out the $395 member price, Ken Magill's article from *Direct Magazine* is a useful source.

When it comes to email marketing, following the best advice will lead you to brand building success. You want people to see your emails, read your emails, forward your emails, visit your website or offline store because of your emails, purchase because of your emails, and most importantly, become brand loyalists.

Unfortunately, without a well thought, preplanned approach to email marketing, you can find yourself labeled as spammer and create a negative image of your brand. Email is one area that can make or break your brand's image, so don't jump in without a great strategy.

CHAPTER 7
THE CIRCLE THAT FEEDS ITSELF

*T*here's nothing more powerful than getting your brand in front of potential customers' faces, literally.

While email marketing messages, Internet advertisements, traditional media advertisements, and other ways of getting people to see your brand are all critically important, one of the most effective means for getting people to connect with a brand is to get them to connect on a personal level. This can only be done one way: **networking**.

Meeting people face to face, shaking hands, enjoying food and drink together, and having a conversation utilizes all five senses. Networking is the only method that effectively allows your brand to touch, taste, feel, hear and see it's prospects.

Via networking, you can establish a brand connection with your potential customers that is stronger and longer lasting than any other method including those of your competitors.

Networking, at its finest, comes in many forms. It could be as simple as running into someone at the bookstore and striking up a conversation where the topic of your brand comes to the surface. It can be a meeting of your local chamber of commerce where the purpose is to mix and mingle with other business owners. It can be an introduction to a mutual acquaintance over breakfast, lunch, or dinner. But remember, the branding cannot stop there.

Before you even begin networking, you must create a networking plan of action and, more specifically, a follow-up branding plan of action. Start by making sure you have all the tools of the trade, such as:

1. Plenty of business cards that properly represent your company and personal brand

2. An opening line that's delivered consistently every time you say it

3. A plan to get more face time, such as attending the next event

4. A comfortable business suit

5. A follow up strategy with the people you meet and from whom you collect business cards

It's true that social media and the Internet have made the exchange of data via a small paper card obsolete. However, there's still no match for the in-person exchange of business cards.

You'd be surprised how many brand impressions the little exercise of exchanging business cards can generate. There's a direct brand impression on that first exchange, but that's only the beginning. Other impressions include situations when that card is lost and suddenly found three weeks later in your jacket pocket, or when that card is shuffled around because it wound up on your desk and you're doing spring cleaning and organizing, or when that business card is handed off to an administrator to be added to a database, or when that card is passed along to another person as a referral. And if you're anything like me, you hold on to that business card for some unknown reason (pack-rat-itis) forever.

Brochures, postcards and other materials get thrown away quickly. The business card exchange creates a constant cycle of increasing brand impressions via that simple person-to-person exchange.

While we are on the subject, I will give you one tip I use; I always ask someone for their business card, but I don't give them mine unless they specifically ask. This may sound counter-intuitive to taking advantage of all those great business card branding opportunities, but I'll explain my reasoning.

If you've ever been to a networking event and run into those "card pushers" who are practically throwing their cards in your face, you are likely not going to feel too bad tossing that card in the garbage. However, if you specifically asked for someone's business card, that's where the strong personal connection is established.

If people aren't asking for your business card, then what do you do? The simplest solution is to ask them for their card. It's a good ice breaker, gives you a chance to study their name, title, and company, and provides a basis to start a discussion. At least half the time you ask for someone's business card, they will ask for yours in return. Not everyone does ask in return right away, but my experience shows that before the end of the conversation, I'm usually asked for my card.

When you are asked to provide them this brand impression opportunity under that person's own terms, your brand has much more impact, and you increase the likelihood they will recall it in the future. What drives me crazy is the number of people I meet during networking (I can meet and personally exchange 100 plus business cards in a busy week) who either don't have business cards on them or look like you are inconveniencing them by asking for their business card. Those are people you absolutely don't want representing your brand, ever! The worst offenders are so called "big-wigs" who think they are too good to provide their contact information to some small business guy because he couldn't possibly offer me anything of value to my large business.

Business cards are meant to be given out. They are not going to increase in value by hoarding them in mint condition. If you are asked, take the golden opportunity to provide a brand impression to a willing recipient even if your card has out of date information.

This exact situation came up at a trade show/conference. I was manning our exhibitor booth at a trade show and greeted a gentlemen as he walked over. He stopped to chat for a brief moment, and I, of course, asked him his name, if he was with a company exhibiting at the show, and if he had a business card. He professed to be the CEO of the company with the very large and expensive looking booth across the convention hall. He mentioned that the business card he had on him had his London address, and that their main operations are in San Francisco, and he's just traveling. I'm taking that as an excuse that he's too big to be wasting a business card with the small business guy at this booth. As you might assume, he did not give me this highly prized business card.

The ironic part is that I cannot remember this person's name or company affiliation. There was an opportunity for him to provide a lasting brand impression, but he failed miserably because his London cards were simply too valuable to give away! Should I later be able to remember, I will remember this as the company with a brand that represents excuses from the highest level of management. Not exactly a company I would want to refer or do business with. And by the way, I know people from large and small businesses alike who could be potential customers.

Again, the simplest of things, such as the business card exchange, can expose a lot about a brand. My experience clearly shows that networking is absolutely not a one-to-one relationship. If you are simply sizing up the potential value that a single person might deliver, you are missing the big picture of networking.

It's not what that person may be able to purchase from you; it's who that person knows who can purchase from you. Networking is truly a one-to-many relationship because you never know who else a person knows. You don't know everyone in their network, and you don't know what influence they may have over those who could be in a position to select your brand as the brand of choice. You need to treat each personal brand impression as the one that can lead to great things, because eventually it may.

In fact, networking can lead to things you may have never expected. In the following blog example, I describe how deep networking can go and some of the pleasant surprises it may bring.

Your Website Can Lead to Engagement (Literally)

http://www.bannerview.com/banner-blogs/mark-cenicola/story.bv?storyid=1315

Trying to find that special someone to ask for your hand in marriage? Ever thought of using a popular online dating site to help you find love online? Ever think that your own website could be the answer? It may sound far fetched as to how your business website could lead to an actual engagement, but just that actually happened at BannerView. com. No, no one contacted us via the website and asked to marry us, but indirectly the BannerView.com website played a role in a recent employee engagement.

A couple of months ago, our Vice President, Jeff Helvin, proposed to his girlfriend. No, he didn't do it online, but of course, reenacted the engagement in front of the BV Ride! Indirectly, however, the BannerView.com website played a roll in his engagement. Here's how:

About 6 years ago, we built a website for a customer. That customer promoted their website address on their company van. Someone saw that website address and visited it. In turn, that person saw that BannerView.com built the customer's website and visited our website. They inquired about openings with the company, ended up interviewing with us and was hired. That employee then built a relationship with and brought on a new customer for whom we built another website. That customer in turn referred us to another customer for whom we built a few websites. That customer had an employee whom Jeff met at a party thrown by the customer who turned out to later become his girlfriend and now fiancee.

While there were a lot of moving parts for Jeff and his fiancee to get where they are today, without our website or any of the customer websites in between, it would have likely never happened! The Web, as it's rightfully called, is a complex medium that can lead you to something you would have never imagined. The moral of the story is that there are many ways to look at your website and what it can bring, not only professionally, but also personally. It took 6 years for this series of events to occur so it's also important to not look at what your website is going to provide today, but what long term prospects it may bring.

One thing that's going to be important, especially if you aren't necessarily comfortable meeting strangers, is an opening line. The opening line should contain only the basics: your first name (last name optional), and your company name. For example, I'll approach someone and say, "Hello, my name is Mark with BannerView.com."

However, not only am I consistent in what I say, but also consistent in how I say it. I was at an event and a fellow associate introduced me to another person at that event. I introduced myself and after doing so the associate commented to me that I always say BannerView.com the same way every time. I chuckled because I didn't notice that I naturally said it with the same inflection, tone, and volume until he pointed it out. It made me realize how comfortable I have become introducing myself, and that I was giving each person a consistent message.

Since branding is about knowing what to expect, people who hear me introduce myself to another person come to rely on my consistency. Whether they realize it or not, that consistency is a reflection on my personal brand and the brand I represent. If you feel the need to introduce yourself differently each time, not only are you going to do it less naturally, but it will send mixed messages by being inconsistent.

Once you are past the basic introduction, the natural next step is for the other person to introduce themselves and to tell you their name, their company name, and/or ask you what your company does. That's when you better be ready to respond with a quick sizzler. Resist the urge to give them your life story or try to tell them every single thing your company does. At most, they will pay attention for 5 to 15 seconds because they are busy figuring out how they are going to respond.

In this case, when I'm asked what my company does, I say, "BannerView.com builds great websites that earn recognition, respect, and rewards for our small business customers." Most people are wowed by my response since they aren't used to a consistently delivered message that gets straight to the point.

Now, I wasn't always this smooth, and neither will you be at first. It took practice, but what has really made it easy for me is that we rewrote our mission statement to be this precise. Telling people what we do is simply repeating our company's mission.

This means you should have a company mission that represents what you are trying to help your customers do. If you work for a company that has a three paragraph mission statement, you'll need to condense it down so you can repeat it consistently every time in 5 to 15 seconds.

If you are the owner of such company, change it now! If people in your organization cannot repeat your mission in 5 to 15 seconds, they will not only have a hard time understanding it, they will also not be able to repeat it in a networking situation.

What comes after you deliver your company's mission is the opportunity for further discussion. You don't have to worry once you engage in a conversation because you'll have plenty of time to get into the details of what you do. Where you go after your opening line and mission statement delivery doesn't have to be as precise or scripted.

Try to keep in mind that networking is simply one aspect to the branding equation and it doesn't stand alone. When you first meet someone, you are skeptical of them, and they are skeptical of you. People are generally skeptical of strangers and prefer to trust friends.

So how can networking position your brand as the one to trust? It's a simple numbers game. You meet someone, exchange cards with them, add them to your email list, and try to catch them at another networking event. As the number of brand impressions to which they get exposed increases, the more likely they are to understand what your brand represents and make educated decisions to whom (if not themselves) your brand speaks. It takes time to build that trust and develop close acquaintances. Supplement your personal brand impression with other brand impressions to build that trust.

Networking regularly is key to building lasting brand value. It you simply attend one event, move to the next event, and so on, you'll never have a chance to build lasting relationships and establish your brand consistently in the mind of people. It's important to realize that branding is about repetition.

Someone once told me that it takes at least three encounters before someone will become comfortable with you. If that's true, you'll need to attend a particular event at least three times before people will become comfortable with your personal brand. Not all networking opportunities come in threes, but you'll find that if you are a regular networker, you'll run into familiar faces quite often.

When people see you out and about, it's best to present yourself in a professional and consistent manner. If people see you at one event in blue jeans and a t-shirt and in a full suit and tie at another event, you'll give mixed branding messages. It's widely accepted to wear a suit at most business networking functions, and if you aren't sure that it's a formal event, you can always wear a blazer or suit jacket without a tie. That will give you a professional, yet acceptable, appearance even in a casual setting. But remember, it's always best to be overdressed than underdressed.

I personally get a lot of compliments on my attire because I strive to always be seen wearing a suit and tie. People are comfortable seeing me in this light, and it is a direct reflection of the brand I represent.

By establishing your brand early and maintaining your first impression, people will hardly recognize when you aren't at your best. It's easy to tell something's up if you don't see me in a suit at a business networking event. It would be out of the ordinary, and you'd probably be right!

Looking the Part Plays a Vital Role in Developing a Successful Online Brand

http://www.bannerview.com/banner-blogs/mark-cenicola/story.bv?storyid=1383

Just recently, at a Las Vegas Chamber of Commerce Business Education Series (BES) Event, a fellow BES Committee member commented to me afterward that I was looking dapper as always. Of course, I thanked him for the compliment and walked away. Little did he know that I was feeling pretty miserable that day as I was suffering the effects of one of the worst allergy seasons in Las Vegas. Even though I may not have been at my best, he was none the wiser.

If you've seen me out and about mixing and mingling in the business networking world, you too probably have noticed that I usually dress consistently by wearing a suit and tie. What many people may not realize is that I'm actually a pretty casual person and don't have to be wearing a suit and tie to be comfortable. However, I do like to present an image of professionalism. By maintaining a consistent look, even when I'm having an off day, most people won't notice. By presenting a consistent look that people are familiar with, and even in some cases appreciate, my personal brand is easily recognizable.

This concept of looking the part plays a vital role in building and maintaining a successful online brand. Every business has their challenges, but by maintaining that familiar presentation style to the world, even when your people are having a bad day or your bottom line is having a bad quarter or year, people are none the wiser. Having been running a business for more than 10 years, we've seen our fair share of ups and downs. In fact, some of our best marketing efforts have been executed at the worst of times and some of our worst marketing efforts have been executed during the best of times. During both ups or downs, we've strived to maintain a positive appearance both online and off. That not only means keeping our place of business tidy, but vitally as important, maintaining our brand online such as our website.

Could you imagine what would happen if you failed to maintain your place of business during both good and bad times? Would you not change burned out lightbulbs? Would you not empty the trash? Would you not vacuum the floor? Would you not restock the shelves? Would you stop properly greeting customers? If you've been to a place of business like this, how much confidence does it instill in you that they'll be around the next time you need something? They could be flush with cash and just forgot to maintain a few things. Whereas another business has a great storefront and presentation, but could be suffering dearly. You are never going to really know what's going on behind the scenes, but 9 times out of 10 you will probably spend money with the company that appears most successful.

Having a bad day is no excuse to dress down or appear unkempt just as it isn't an excuse for failure to maintain your online brand. That dollar you earn may be just the break you need to take your business to the next level or that compliment you earned may be just the positivity you needed to turn that bad day around!

It's what you do after the networking that's most important. One of the often heard comments I get when networking is "I see you everywhere!" The obvious conclusion would be that if they see me everywhere, then they must also be everywhere too! That is certainly not the case. The reason I get that comment is because of all the other branding techniques we combine together to make my personal networking so much more successful. In fact, it creates a circle of branding power that constantly feeds on itself. In some sense, "I see you everywhere," is true because I don't stop getting in front of a person when we are no longer face to face.

Having a strategy of follow up before you hit the networking circuit will lead to others telling you, "I see you everywhere!" While your follow up strategy may differ, here is what has been successful for me. As previously mentioned, I may meet and exchange business cards with more than 100 people in a week. It's not really feasible for me to personally follow up with every single person, nor would I necessarily want to do that. However, I can add them to an email marketing list and follow up on a consistent basis. That email is going to be properly branded, have valuable information that's relevant, display my photo, and have information that's personally authored by me.

Many networking events also have people who take photos and post them on the event's website. Here's not the time to be camera shy. Get in those photos. All the regulars who couldn't attend will probably be looking at the photos. When they see you, they'll associate your brand as being there even though they weren't!

It also helps to have multiple people from your organization network. They may meet someone you didn't. However, the new connections will receive the same email communication that contains your name and face. Therefore, when they do finally meet you, it will feel like they already know you.

As a small business with a limited marketing budget, the chances are that networking may be a primary means of branding your company. There are plenty of books providing tips and advice on networking, how to go about it, and how to build great relationships. The purpose of this book it not necessarily to try to coach you on the proper networking techniques, but rather to expose you to the importance of networking when it comes to building your brand. Take the time to study networking, but look at it from a branding perspective, which is a point of view few networking advice books take.

Networking has been looked at as a way to generate new leads quickly. If you do happen to generate a lead via networking at a particular event, great! But looking at networking activity from a branding perspective takes the pressure off you and those representing your brand in public; you don't feel desperate to generate a lead.

In fact, the majority of our business has come as an indirect result of networking. I say "indirect" because we've always supplemented our networking with follow up and brand impressions via different methods. It would have been easy to pinpoint the result of a sale back to meeting someone at a networking event, but without exposing those prospects to other branding impressions, the sale may have never happened or it might have simply been superficial in value.

Focus on networking as a branding strategy and you'll build deeper relationships, attract the right kind of customer, and have loyal customers out there promoting the value of your brand via the wonderful stories about their experience with your organization. Combine networking into an overall branding strategy, and you can leverage contacts to include introductions to others outside your circle of influence. Soon you will have literally hundreds of people acting as brand ambassadors on your behalf when you aren't physically present.

CHAPTER 8
BOOK TO BRAND

*A*re you kidding? Me, write a book? That's what I would have said two years before actually writing this book. How the heck do people have the time or the patience to sit down and write books?

Well, for me, it's called leverage.

Many people see the value of a book, even if they are not writers, or don't have the time or inclination to write. For these people, ghostwriters make books possible.

Guys like Jonathan Peters are hired to write books for others. Jonathan gets paid for what he does best—write! The person with the name on the book gets all the credibility for the work of others.

My solution was a more middle ground. Leveraging all the resources, knowledge, and stories over the years provided the perfect opportunity to make the book writing process much less time consuming and easier to tackle. The chance to use previous material for covering certain topics, and the practice and consistency of sticking to a weekly publishing schedule on my company blog, gave me the leverage necessary to make writing a book a reality.

At the same time, I had someone with Jonathan's experience to guide me through the process. No, he didn't actually write this book; I did. Please believe me! I'm sure he wants you to believe me too.

I don't recommend hiring a ghostwriter (but if you do, of course, hire Jonathan Peters). When you know your subject well enough to write a book, your speaking gigs and other publicity efforts will be easier to handle. Plus, having done some of the legwork will show your sincerity. After all, credibility is certainly a key to having a successful brand.

Writing a book is a great way to strengthen not only your personal brand, but also the brand you represent. That's why companies hire star athletes. People buy into the credibility of the superstar and associate it with the advertiser's brand; at least that's the hope.

In my case, writing this book is simply a way to add to my credibility so people have more reasons to listen when I'm representing my company's brand. Plus, this book was another avenue to get in front of people, let them see me, and the brand I represent with the intention of building a personal connection. My ulterior motive is to get them to purchase the services from the brand I represent. Again, leverage is in play here. Take what I'm already doing, find ways to use it to the brand's advantage, which will result in increased sales for that brand, and voila!

As soon as people find out you're a published author, their attention focuses a little more closely on what you have to say. Regardless of the fact that you may be the only person to ever read my book, I still have credibility for all the people who haven't (yet) read this book.

If you still aren't convinced that you have the where-withal of time, patience, skill, etc. to write your own book, even with help, there's a good alternative that can be found: Hire an author! What I mean is you can hire someone to work within your organization who's already a published author and make them a brand ambassador. Just be sure that they are an accomplished public speaker or have the potential to be. Someone who is charismatic and well liked by audiences is most important, even more so than the quality of their written works.

While we were in the process of searching for people to expand our executive team, we came across a person we felt would be a good fit. This person was also a pub-lished author. Since I already had many connections with various local organizations, I knew I could leverage this person's author credibility to get speaking engagements. Knowing full well that those speaking engagements would create brand exposure, I thought we might have a winning combination.

In the end, we couldn't come to terms with this person during employment negotiations, but it certainly had me looking at this possibility with future executive hires.

That exercise opened my eyes to the fact that a person's published work(s) can be a gateway to get them speaking opportunities and other appearances that would give our brand more exposure. While they would need to leverage their works, and we'd have to be okay with them selling their books, having an integrated strategy for them to promote their published works could have given our brand the exposure it needed.

If you do decide to go this route, make sure that their employment within your organization is going to be a backdrop in which you can create a win-win situation for both you and your employee. They get to promote and sell their book, and you get the brand exposure via the affiliation with the published author.

In the end, it made more sense for me to write my own book and leverage that. However, for those of you not as ambitious to go the full blown book authoring route, there's leveraging your blog to line up speaking gigs. For a case in point, read the blog entries below:

The BANNER BLOG

Lining up the Speaking Gigs

http://www.bannerview.com/banner-blogs/mark-cenicola/story.bv?storyid=1242

June is going to be a busy month in Las Vegas for presentations. I'll be presenting at 3 different functions on the following topics:

Effective Use of Blogging to Drive Website Traffic

Searching for Seniors Using Email Marketing and Blogging

Using Email Marketing & Blogging to Sell your Personal Brand

As previously mentioned in my blog, I've been invited to speak as one of the 6 presenters in IABC Las Vegas' 6 in 60 program on June 2, 2009 at 11:30am at Maggiano's Little Italy inside the Fashion Show Mall. Come join me as I present to professional business communicators the right ways to utilize blogging to drive traffic to their Websites.

I've been invited to present to S.I.N.G. (Senior Industry Network Group) to talk about how blogging and email marketing are effective ways to reach those targeting the senior demographic. This event, located at Marie Calendar's on Flamingo and Decatur, will begin at 11:30am on June 11, 2009. This group is made up of professionals that target the senior demographic.

I've been invited to present to the Nevada Professional Coaches Association to discuss how blogging and email marketing can position these professionals as experts in their field and keep their person brands in front of prospects. This event will be held at McCormick & Schmick's on June 19, 2009 starting at 11:30am. This group is comprised of both life and business coaches who help their clients achieve their personal and professional goals.

Come join me at all three events to learn more about how to effectively use email marketing and blogging to generate business and network with other Las Vegas professionals.

If you are interested in having me present to your organization or group in the field of e-business solutions, feel free to contact me via our online form.

The Chicago Bust Out speaking engagement lead to the next blog article, which shows how I leveraged that to produce more material for my blog which is ultimately being incorporated into this book.

The BANNER BLOG

Speakers, Trainers & Coaches Geared up to Bust Out!

http://www.bannerview.com/banner-blogs/mark-cenicola/story.bv?storyid=1320

This past weekend, I had the pleasure of spending the weekend in the Chicago area for the Chicago Bust Out Bootcamp. As previously mentioned, the program was geared specifically toward professional Speakers, Trainers and Coaches who were interested in venturing into business on their own instead of relying on booking agencies and nationwide seminar companies to generate their income. That's a perfect audience to address as we have a number of coaches and speakers that have been successful at building their own personal brands by utilizing the multitude of eBusiness products that BannerView.com has to offer.

The entrepreneurial spirit was alive and kicking and you could feel the excitement from the participants as Monica Cornetti from EntrepreNow kicked off the program on Saturday morning. That afternoon I presented the Top 5 Tips Every Small Business Owner Should Know About SEO. While it's a very technical topic, the feedback I received assured me that I was able successfully take what can be a dry topic and make it interesting for everyone to understand and stay awake, especially just after lunch! SEO, or Search Engine Optimization is best handled by professionals that specialize in that area, but it's very important for entrepreneurs looking to start or expand their business to understand basic concepts and know what questions to ask their web developers.

The second day of the program (Sunday), I kicked off the first presentation of the day with the topic Using Email Marketing and Blogging to Sell your Personal Brand. Thankfully, everyone was wide awake and ready to kick off the morning. The audience was engaged with many questions, laughter and excitement about the opportunities available to them to promote their brands.

While focused on delivering my programs as a featured presenter, I still came away with many ideas from listening to the program organizers, other presenters and audience members. In fact, I even learned from myself after providing a tip on how to use the unsubscribe page on your email newsletters as a lead capture tool, something we still needed to put into use. If you have an email newsletter with great content that provides value to your visitors and is consistent with all of your existing branding, it's not likely that people will unsubscribe. However, in the unlikely event that a subscriber wishes to leave your list, you can inform them of the many benefits they'll lose by unsubscribing and provide a cancellation button on your unsubscribe page.

The program offered a lot of useful tips and advice that can immediately be placed into action. My presentations were interactive and well received. We generated interest in the products and services that can help the target audience implement to their goals and the weather in Chicago was beautiful with highs in the upper 60's, in November, go figure!

I continued to build off my previous speaking engagements and my blog. This next blog article shows that the speaking engagements continued even without a book to backup my credibility. I simply used previous business success and speaking success and took my show on the road.

The
BANNER
BLOG

Educating the Windy City at the Chicago Bust Out Boot Camp

http://www.bannerview.com/banner-blogs/mark-cenicola/story.bv?storyid=1303

On the heels of multiple successful speaking gigs in Las Vegas and in neighboring Phoenix, it's time to jump on a plane and head to the Windy City for the Chicago Bust Out Boot Camp November 6-8, 2009. This program, designed to enable **speakers, trainers** and **coaches** to "bust out" from:

- relying on one or two clients for the majority of their revenue.

- allowing a public seminar company to control their calendars.
- using a stale business model that just isn't working anymore to build their own brand.

With an extensive program full of quality speakers I'm exited just to be able to attend. However, I'm just as exited to be personally presenting two sessions that are vital to promoting a successful online presence:

- **Effective Use of Blogging & Email Marketing to Sell Your Brand**
- **The 5 Things Every Small Business Owner Should Know About SEO**

One of my fellow presenters at this seminar will be Jonathan Peters Ph.D., Las Vegan, UNLV professor, author and owner of Circumference Communications. Of course, credit goes to him for helping line up this gig for me as a friend and customer of BannerView.com. It resulted from a conversation we had during his attendance at the latest BannerEdutainment Tour which is further evidence our own email marketing and blogging efforts continue to pay dividends in ways not originally envisioned.

If you plan to be in the Chicago area in November, check out the Bust Out Boot Camp as it's sure to be an exiting and eventful program.

Don't wait to line up the speaking engagements until after you've written your book. Start the book and use your experiences to provide ideas for that book. Then, once the book is written, revisit all of the previous engagements as a follow up opportunity to further validate what you presented, and promote the book!

CHAPTER 9
THE GLUE THAT TIES IT ALL TOGETHER

Your website is the most critical component of launching a successful branding campaign. It's the foundation on which to build your brand, and it is the glue that holds all elements of your branding campaign together.

Before the Internet, a business' physical location was where its brand rested. It was the central point where business was conducted, the place where products originated, home to its employees, and the place where a customer could experience the brand.

The Internet and, more specifically, the World Wide Web, has extended a business' brand outside of its physical location. Through a website, customers can now get intimate with a brand and build loyalty without ever visiting a physical location.

However, it's important to note that the Web did not replace the physical operating location. That place still serves as the place where employees go to work (telecommuting was supposed to have eliminated that), the place where products originate, and it still provides a strong brand presence in the physical world.

Therefore, it's important to note that even with the prolific rise of social media popularity, the website still serves as a brand's home on the Internet. It's the place where the business can control what gets said, what gets done, with whom interaction takes place, and the central point where the brand's message can be consistently represented.

Social media is simply another extension of a brand's interaction with customers. It's similar to having a distribution channel, like how a soft drink maker relies on restaurants and grocery stores to carry its product to customers. Without a home base, products don't get created or delivered to where they can be purchased.

While you may no longer have control over the way your brand is perceived in the marketplace, or what's being said about it through social media, your website is the place where customers will ultimately go in order to see what's true.

One example of this is Apple. Apple manufactures computers, phones, music players, and related software over which people clamor. The company is probably one of the most followed brands online with a large number of independent websites that report on rumors about the latest products the company is set to announce. While all of this speculation runs rampant, there's only one source that is the de-facto verification point: Apple's website. If it hasn't been posted on apple.com, then it's not verifiable.

On the day that Apple announces new products, I usually receive information reported via the standard news outlets. I then visit the Apple site to get verification of what I read and to learn more.

Many people are now relying on social media to get updated on new product announcements and immediate breaking news. But the need to verify that information doesn't stop with Twitter or some other social media. We must go to trusted sources of information, in most cases the only trusted source of information is the company's website.

In case you aren't yet familiar with the phenomenon called Twitter, it was rumored to be a billion dollar company before most people even knew what the heck it was. I brought this to the attention of my blog readers in the following raw example.

The
BANNER
BLOG

Not Familiar with Twitter Yet? It's a Billion Dollar Company!

http://www.bannerview.com/banner-blogs/mark-cenicola/story.bv?storyid=1218

Coming off our latest E-Business Seminar in Las Vegas where we covered the basics of social media and specifically, Twitter, we discussed how this service can be used to benefit your business by quickly disseminating information to your followers. However, it'll be interesting to see what's going to happen with Twitter now that it might be bought. The rumors are that Google is in talks to acquire Twitter. Will it be purchased for $1 Billion? Who knows, but there's certainly a lot of discussion as you can see in this video from Tech Ticker.

That's a lot of money for a service that doesn't make any money. But hey, Google is the big guy on the block and can afford to *drop* $1 Billion and not even blink an eye. It purchased the money losing YouTube.com for $1.65 Billion and still hasn't monetized the service to the extent that it would like. Maybe we are back to capturing eyeballs on the Web instead of capturing people's wallets. When you have the buying power of Google and can afford to purchase loss leaders to indirectly support your primary business, why not? Unfortunately, not all of us have that luxury so we must be more nimble when it comes to our own Web Site operations. The good news is that there are options available to implement all the tools necessary for a successful Web Site presence at an affordable price.

As of the writing of this book, Twitter continues to remain an independent company. An interesting trend is happening with Twitter that is relevant to why websites are still a central ingredient to holding a brand's image consistent, many Twitter users utilize the service by means other than the company's website.

I wrote the following blog article as an example of how many brands are wrongly investing branding dollars into social media, but overlooking what's still most important. Social media is part of an overall branding strategy, not the starting point.

Web 2.0 - Are You Giving up too Much Control to It?

http://www.bannerview.com/banner-blogs/mark-cenicola/story.bv?storyid=1237

With everyone jumping on the Web 2.0 and social media bandwagon, how many companies now advertise their Web presences via their Facebook, Myspace and Twitter handles? How important is a company like Google that is responsible for more than 60% of the search traffic that drives visitors to a Web site or eBay in which thousands of small business owners rely upon for selling their goods online? If any of them suffer an outage, online activity can come screeching to a halt. Just look at the recent example of Google suffering an issue for only a few hours. Web traffic in North America alone dropped off a cliff and some websites were rendered completely useless as pages wouldn't display due to Google Analytics not loading. How much control do you have over the policies they put in place that could affect your business, the reliability of the service they provide or for that matter, any say in anything that they do?

Before presenting the moral of the story, let's first under-stand a little history of the Internet and its place in the world.

The predecessor to the Internet referred to as ARPANET (Advanced Research Projects Agency Network) grew out of a United States Government project under DARPA (Defense Advanced Research Projects Agency) as a way to link researchers to powerful computers when they were geographically separated from them. As the Internet developed into the resource we know today, it has required the participation of many governments, educational institutions and private businesses to thrive as an open platform of communication the world relies upon so dearly. Many neutral organizations have been created to help regulate and keep the Internet free to end users such as ARIN (American Registry for Internet Numbers) which has similar organizations for different geographic locales outside of the U.S. and ICANN (Internet Corporation for Assigned Names and Numbers). These two organizations control how IP addresses and domain names are regulated providing a fair playing field for anyone interested in establishing their presence online. Numerous commercial organizations are in existence today that can help you establish your presence online. Should you run into a challenge with one of them, you can switch to a myriad of others. All in all, the destiny of your Web presence is mainly in your hands and the Internet presents a pretty level playing field for all to compete (promotional budget disparities aside).

While the social networks, search engines and commerce providers are typically governed by the land in which they operate, there are no neutral organizations regulating how they run their businesses. When a commercial organization gets too big governments tend to like to regulate them. If there isn't enough regulation, they could become "too big to fail" and be deemed a systemic risk to the system, attracting the scrutiny of governments after it's too late. When capitalism comes into play, what was great when it didn't need to make money, now isn't so great when the service is no longer free. Or, when one company gets gobbled up by another company it can completely change the dynamic of that service sometimes making it not so great for your Web presence.

If you put too much reliance for the success of your Web presence in any one of the search engines, social networks or commerce networks, you are at their mercy whenever there's an outage, change in terms of service or government imposed regulation upon them.

The lesson here is to develop your own online presence that's completely independent of the too big to fail companies so you're in control of your own online success. Don't put all your bytes in one medium. Use all available resources to help grow your online presence, but it's better to be prepared to survive without one of the big Web companies.

Here's one more example from my blog regarding the dangers of investing solely in social media while ignoring the central location of your online brand. Just as physical locations continue to exist in the real world, I strongly feel that websites will continue to survive the test of time as the central location for a brand's image.

Let the Social Media Shakeout Begin

http://www.bannerview.com/banner-blogs/mark-cenicola/story.bv?storyid=1249

It was only a matter of time before some of the large so-cial media sites started tightening their belts as evidenced with MySpace announcing a nearly 30% reduction in staff. Maybe it's simply the competition from Facebook digging into MySpace's user base or something more grand in that, despite all the hype surrounding social media, does it actually contribute to a company's bottom line?

One of the Web 1.0 social media sites, Geocities, which was acquired by Yahoo! back in the ancient social Web year of 1999 for over $3 billion, saw its fate sealed in April 2009 with Yahoo! announcing that it was closing Geocitities. If an Internet pioneer like Yahoo! can't figure out a formula to making money from giving services away for free, where does that leave the social media mega-sites?

During a recent meeting with Brian Mell, BannerView.com's Marketing Coordinator, we discussed the true relevance of sites like Facebook and Twitter. Even though BannerView has profiles on both, we still don't know the magic formula for effectively capitalizing them to benefit our bottom line. I guess we still have quite a lot of work to do or the ultra hype in the Web 2.0 space is simply the second Internet bubble readying to burst. We looked at ways to increase our followers significantly on Twitter which we think we'd be pretty successful doing, but what would that get us? From a business standpoint, anything you do that doesn't contribute to profits is wasted energy. The saying, "Doing something that doesn't make money is simply a hobby," regardless of my poor rendition of the actual quote, is starting to ring true with a lot of the social media sites. If the sites themselves can't make money, what's that leave for business folk like us?

I'm not an anti social media/Web 2.0 naysayer, but I have to look at the return on investment. Having people spend all day working the social Web can be an expensive proposition in terms of human capital. While we've just dipped our toes into the water, we are experiencing success via Blogging and email marketing. As those efforts continue to pay off, you can be certain this is where our focus will remain. Grab some popcorn, or a nice long book to read, sit back and watch or wait as the social media market begins its shakeout.

Don't get me wrong, social media is something that can't be ignored. It's important to have a strategy to utilize it, just like any of the ideas presented in this book. In the following blog article, I wrote about this point exactly.

The BANNER BLOG

Strategic Social Media Planning
http://www.bannerview.com/banner-blogs/mark-cenicola/story.bv?storyid=1317

We've covered social media in a number of articles in the BannerBlog, but we haven't really focused on implementing a strategic plan to get the most out of your social media efforts. Quite frankly, it's because we don't have a strategic plan in place ourselves! So how can we advise on social media strategies if we don't have our own? Well, the very simple fact that we don't have a strategy is a strategy in itself. We do have Facebook and Twitter accounts and we do post that fact on our Websites, but never have we formally implemented a plan to meet some sort of objectives or set goals on what we'd like to accomplish with social media websites. Hence, that's the reason we haven't put a lot of effort into social media. Which brings me to the point that since we haven't put together a strategic social media marketing strategy, it doesn't make any sense for us to expunge a lot of resources on it and neither should you!

While having a strategic plan in place makes sense for any advertising and marketing campaign, it seems that because of the buzz surrounding social media, companies are jumping in with two feet before they even understand who, what, where, why and how they are going to measure results. I'm not saying that you can't be effective with social media, but rather, you need to have a strategic marketing plan in place to take advantage of social media.

The BannerBlog and our email newsletter, The BV Buzz, are currently our primary focus. We've strategized on what we want to accomplish with them and have put a plan in place to meet those objectives. We have the resources, tools and people in place to meet our goals within the timeframe we want. Therefore, it makes sense we focus our attention on Blogging and Email Marketing. As a small business with limited resources, we can only get involved in so many things. Until we've strategically planned our social media efforts, we'll continue to stay focused on what we are doing instead of getting caught up in the next Internet craze. Should social media prove to be an avenue of successfully growing our business, then we'll put a strategic plan in place to make it happen.

Finally, here is the last blog post about social media that I'll include in this book. It's an interesting take on how social media is a roundabout way of building your brand.

A Complex Formula for Finding Me Where I Work!

http://www.bannerview.com/banner-blogs/mark-cenicola/story.bv?storyid=1283

Warning, this blog entry contains complex hypothetical mathematical formulas and assumptions that have not been tested by any known officially qualified personnel or government institutions. Proceed at your own Risk!

Many companies are going out of their way to promote the fact that they are using social media. You'll see ads promoting to find them on Facebook, follow them on Twitter, view their Myspace page and link up with them at LinkedIn. Whatever happened to good old fashioned, "Find us at our home, on our own website?"

When's the last time you spent money advertising for a potential customer to come meet your company at the gym, at Starbucks or at the mall (unless your location is actually the mall) before you've even pre-qualified them? I wrote about giving up too much control to Web 2.0 in May, but I just saw an ad in the back of a magazine promoting readers to visit that company on Facebook that inspired this rant. It seems to me like saying, "I hang out at a cool place so I must be cool and maybe you'll find me cool because I hang out at cool places." Sound a little desperate? Would you pay for ad space that directed people to find you on Yahoo!, Google or Bing? Wait, is this a trick question? The correct answer is No, because you'd pay Yahoo!, Google or Bing directly to help send potential customers your way. Spending money to promote your business on a social media site is equivalent to spending money to promote the fact you can be found on Google, then paying again to have Google finally send the people to your website after they searched for you on Google.

This all seems way too complex. Let's look at it as a mathematical formula.

You want to be found, so you pay Google, for inclusion in their Adwords pay per click program, to send visitors to your website. Google is "G" and your website visitors are "W." This provides a pretty simple formula.

G + $ = W

Now let's say you pay money to someone else (like a magazine) to advertise the fact that you can be found on

Google. The Magazine is "M" (substitute any advertising method for M) The formula now becomes a little more complex like so:

M + G + $ = W

So you argue that social media sites like Facebook are free and if you are listed on Facebook, you don't pay any money. That formula would look like this. "F" is for visitors to Facebook (substitute your favorite social media website for F).

M + $ = F

More than likely, not all visitors to Facebook would visit your website and the ultimate goal is still to get people to your website or storefront in order to make a purchase. We'll call that the X-Factor or "X." So your formula looks like this:

(M + $)/X = W

Oh, and you realize Facebook isn't free after all. It does cost money to maintain a Facebook profile since time is involved whether you have dedicated staff or you take the time to do it yourself. We'll call that "S" which gives you the following formula:

(M + $)/X + S = W

Wow, all these formulas are starting to get complex! So let's simplify again and go back to the magazine ad scenario and just advertise your website directly. That provides a simple formula of:

M + $ = W

Out of all the formulas, whether you choose **G + $ = W** or **M + $ = W** (feel free to substitute G & M for any advertising medium) as your preference, these are not only much simpler formulas, but are also going to be less costly.

Sometimes it's okay to try formulas that are a little more complex as you get more sophisticated as they may pay off. However, for most small to medium sized businesses, resources aren't available to calculate such complex formulas. Therefore, if you find yourself comparing options, the clear conclusion is that the simple choice is usually the best choice when resources aren't abundant.

Well enough on social media. It already gets an inordinate amount of attention, and there are numerous books on the subject. I brought you to this chapter to explain the importance of your website as the glue that holds all of your branding efforts together. I figured it was important to at least provide a cursory overview of social media since it's one of the key aspects of your brand, but it gets more attention than good 'ol websites. The reason is that websites are looked at as "old media" and everyone loves new media.

In fact, there have been a few changes in how websites are built over the last 15 years now that broadband has proliferated in people's homes and businesses. This means that more and more multimedia, such as video and Flash, are getting incorporated into the website experience. This blog entry shows how even we are utilizing more video in our online branding efforts.

The Power of Web Video and Planning for its Success

http://www.bannerview.com/banner-blogs/mark-cenicola/story.bv?storyid=1328

A number of our clients are beginning to understand the power of video. In fact, many are adding video to their websites right on the home page or incorporating video in product demonstrations as part of the sales process or in "how to's" after purchase. With the proliferation of broadband, it's not much of a concern anymore for the end website visitors to watch video as speeds are good. Now, it's more of a concern on the webmaster side on how to effectively implement and use video.

There are really two ways of going about adding video to your website. You can either upload your video to a service like YouTube or host the video directly from your own website. There are advantages and disadvantages to both.

Advantages to using a video hosting service such as YouTube

* No bandwidth needed - this means that the video doesn't use any bandwidth at your website host to maintain the video. YouTube has vast resources available for serving multiple simultaneous views of your video. Your web hosting service may not be geared for video hosting, have a limited number of simultaneous views or cost a lot of money to maintain performance and bandwidth needed should your video go viral.

* Video hosting services have made it easy to embed video within your website content. For example, in our Halloween at BannerView.com video, we have it embedded from You-tube. Scroll to the bottom of the blog entry to watch.

* Youtube provides a comments/feedback system so people can interact by posting comments and starting a dialog based upon the content of your video.

* Youtube includes your video in its search results and Google is now embedding video results from Youtube directly in it's regular web search results.

* Popularity or how many times your video has been viewed is recorded. Popular videos can go viral and your popularity can feed upon itself.

Disadvantages to using a video hosting service such as YouTube

* The video hosting branding is embedded in your video. When you embed a YouTube video, it's pretty clear that you are doing so. This means you lose a little control over how the branding of your video appears and may detract from the display flexibility of the video on your Website.

* YouTube currently limits videos to 10 minutes according

to their help documentation. If you have product walk-throughs or a need for longer videos, YouTube might not be the best option.

* While YouTube is currently experimenting with HD quality video, the quality can suffer during the conversion process. as they've noted in the help documentation

* When YouTube suffers problems your videos don't get displayed.

* People can comment on your video, while listed above as an advantage, you may not want people commenting on a video with bad comments.

* People can see how popular your video is. While an advantage in some situations, people may be reluctant to watch a video if very few others have.

So there are just a few advantages and disadvantages of using a video hosting service when considering the use of video within your website. With the pros and cons of each method, you can start to plan the best method for incorporating video into your website. It appears video is the next wave of Internet content to take the world by storm!

Stay tuned for video to become a part of BannerView. com's Web strategy going forward in 2010. I'll keep you posted on what we learn during our increased use of video online.

The new developments in Web technology have provided additional branding opportunities that were previously not available. In addition, tools like email marketing products, blogging, eCommerce applications, and social media plugins have become affordable. Previously, only large businesses had the resources to enjoy incorporating such elements into the website experience.

However, with the proliferation of all of these tools, many small businesses are implementing them hastily which means that they are actually hurting their brand instead of helping it. There was good reason that only large businesses succeeded in building great brands online. They had the staff, resources, and money to make sure their brand's message didn't get lost between a huge disparity of low cost tools. Instead, they built and continue to build many of their own systems for handling email marketing, blogging, eCommerce, and social media integration.

Let's look at what most businesses want from their online presence:

- They want a great looking website that portrays their brand image

- They want a website optimized for search engines

- They want to incorporate the ability to generate leads or sell products directly via their website

- They want to incorporate social media tools like blogging

- They want email marketing capabilities that keep their brand in front of customers

- They want to be able to maintain their website with up-to-date information

- They want the site to stay accessible to the world 24/7

Here are the mistakes that most businesses with limited resources make when sourcing their online presence:

- They find the lowest cost web designer to design their website
- They hope that the web designer will build them a website that's optimized for search engines since it's too expensive to hire another person to do it right
- They turn to the most affordable third-party eCommerce program to plug into the website
- They turn to free blogging solutions widely available via third-party websites
- They turn to a low-cost, third-party email marketing companies
- They turn to an easy, do-it-yourself web updating tool to make changes
- They turn to the lowest cost web hosting company to host the website

Now here is what most businesses get by going the above route:

- They get a poorly designed website that doesn't properly portray their brand
- Their website is not properly coded, and therefore, has trouble being picked up by the search engines
- They get an eCommerce product that doesn't fully integrate into their website, and they struggle to get it to work properly
- They have a separate website that hosts their blog, which confuses visitors with a separate brand message

- They get an email marketing solution that doesn't convey the company's primary brand and causes confusion
- They struggle to make updates to their own website, and when those updates are made, the pages don't look very professional
- They have downtime, slow performance, and a don't-care attitude from the provider

Now, let's break down each critical component of an online presence that increases brand awareness, provides visitors a seamless brand experience, and contributes to the success of the business brand.

The first component is a great looking website that portrays the brand image. This is where small business "terrorists" get involved in the process. They trick the small business owner into low cost solutions to build a website. The problem is that the website doesn't portray a professional image, and it looks as if an amateur built it (and they probably did)!

If you were building a skyscraper, you'd hire a professional architect to design the building. You'd want to make sure the foundation was properly designed to hold all the stories in the building. Your website, while not as costly, still needs the same care to make sure it's properly designed to stand up to all the elements you want to plug into it.

The second critical component of an online presence is making sure the website is properly coded to be found on the search engines. If you currently have a website, visit http://validator.w3.org/ and type in your full website address. If you get any other result other than a green "Passed" result, your website is not properly coded, and it's costing you business. The following entry from my blog goes into depth on this topic:

Your Website is Not Properly Coded and It's Costing you Business!

http://www.bannerview.com/banner-blogs/mark-cenicola/story.bv?storyid=1334

One of the most overlooked aspects when it comes to building your online presence, is properly coding your website to meet widely accepted standards. While your website might look good on one Web browser, does it look good on 6 other browsers on the multitude of other systems? There is Windows XP, Windows Vista and now Windows 7. Each of those have varying versions of Internet Explorer and Firefox. Of course there is also Mac OS X with different versions of Safari and Firefox. If you've ever visited a website and it doesn't look quite right, such as text overlapping other text or images hiding text, that could be a cross browser compatibility issue.

For the most part, cosmetic issues across different browsers can be normal occurrences and if they don't affect the usability of a website, they are simply minor nuisances. However, the bigger issue is when important functionality such as the ability to purchase products, perform transactions or exchange data, is affected. It's too important to let those things go untested across various platforms as that can lead to lost business. Another big issue is if the search engine robots (spiders) can properly read your website. Those are the little programs that crawl the web looking for content to include in major search engines like Google sends out. If your website isn't properly coded, then you risk not getting found on the search engines which makes successfully conducting business online much more difficult.

One way to see if your website is properly coded to work across web browsers and ensure those search engine robots can crawl your website is to check to see if you are W3C compliant. W3C, which stands for World Wide Web Consortium, is the international community which sets the standards for how the web is supposed to work and how web browsers and websites are supposed to be coded to work across all platforms. Websites that follow W3C standards usually perform better on the search engines when it comes to getting higher rankings. The problem is that most Web development companies don't follow W3C standards when coding websites.

You can check your own website for W3C compliancy by visiting their validation tool. If you find that there are numerous errors and warnings, then optimizing your website's code for W3C compliancy can be a way to improve your chances of success, not only on the search engines, but across browsers and different operating systems. Should your site have errors and warnings, BannerView. com can help with our Website Optimization Service which takes into account properly coding your Website as part of the optimization process.

Why pay an architect to properly design your building, but not have a licensed contractor follow the plans to make sure it's constructed properly? Designing a website and constructing the website are two completely different skill sets. A one-size-fits-all approach to design and construct your website doesn't work. If designing a building was so easy, anyone could be an architect. Well everyone is not. Similarly, designing websites is not easy. I explain this in the following blog entry.

The BANNER BLOG

Designing Great Websites is NOT Easy

http://www.bannerview.com/banner-blogs/mark-cenicola/story.bv?storyid=1243

We recently had a customer go through a complete training class in Dreamweaver so they could take on some of their website work themselves to save some money. We support our clients when they bring things in-house and work with them through the process. After this process, the client profusely recommends we make all of our clients go through a Web design class since they now have a completely new respect for what our company does and thinks everyone needs to know. Of course we are here so everyone doesn't have to go out and become Web designers since business owners should have much more important things to do like running their companies!

It's true that a lot of professions don't get as much credit as they should and some professions get more than they deserve. Most people have no idea what goes into building a great website and no, it's not as easy as it looks. There are many considerations that go into designing successful websites and the actual construction is simply one aspect that must be done properly.

If you think you're going to take the cheap & easy road and have your son's girlfriend's mother's cousin design you a great, effective website, you must ask a few preliminary questions:

"Has that person built other successful websites?"

"Do they or you know what a great website should be?"

The first question may seem easy to answer as in "Yes, they've built so and so's website and it's great!" That's where the second question dives in a little further. Simply having a great looking website, while a much sought after desire from a site owner, isn't the real question that needs to be answered. The best question that needs to be answered is:

"Will the website meet the business objectives?"

In order to answer that question, you first need to know the objectives. Some **website objectives** may include:
- Increasing brand awareness
- Generating leads

- Reducing support costs by bringing solutions to customers online
- Selling products

The business objectives for what you want your website to accomplish need to be answered by the business owner, not determined by the designer. Great web designers will take your business objectives and build a solution to meet your needs at your acceptable budget. There are plenty of so-so Web designers that have constructed websites, even great looking ones too, that never meet the business objectives of their owners. The toughest part is that you can't tell just by looking at a website whether or not it generates revenue, builds brand awareness or saves the site operator money in customer support costs.

Great companies make things look easy and it's a compliment to think we make the development of websites look easy enough that anyone could build a great website! However, the reality is that great websites are made by our customers' visions, clearly defined objectives, hard work and of course, the monetary investment into the solutions needed to be successful online.

The third critical component of an online presence is the ability to generate leads and sell products online. What many businesses don't realize is that when they use third-party eCommerce products, also referred to as shopping carts, many of them are also hosted on third-party websites. This means that people actually have to leave your website to finish the transaction. This is similar to someone walking into your physical store, picking out the products they want to buy, then telling them they have to walk across the street to a kiosk to make the payment. This not only confuses the customer, but has them wondering about the security of their transaction.

Keeping the customer on your website throughout the entire purchase process is the best way to keep a consistent brand image and allay any fears that there could be security risks to doing business with you. It's imperative that your eCommerce system integrates directly with your website to avoid branding inconsistencies and appearances of security risks to the customer.

The fourth critical component of an online presence is having a blog that's fully integrated within your website. Why would you go to a third party and have them host a blog separately from your website? By doing that, you now have to promote two different websites that probably look completely different. Integrate the blog directly into your website and you not only have a single website to promote, but you also ensure that the look and feel of the blog is going to match that of your overall website.

Your blog is a component of your overall online presence, not a separate entity, so don't treat it as such. A properly integrated blog provides valuable content to the website visitor and creates a positive image of your brand.

The fifth critical component of an online presence is having an effective email marketing strategy. This means you want a product that's going to directly integrate with your website so that when people subscribe or unsubscribe, they feel comfortable doing so.

Too many people choose a third-party email marketing service that, when you subscribe, takes you to a blank page, and when you unsubscribe, you are taken to another blank page. This introduces confusion to the customer by causing questions about whether they are subscribing to the correct email list, or are they actually unsubscribing instead of being duped by a spammer.

The unsubscribe process is actually another opportunity to explain to the customer what they will be missing should they proceed with unsubscribing. You can't do that if the page is blank.

Also, when you send out your email communication pieces, it's imperative that they are consistent with your brand. Too many times I receive emails that are poorly branded, and I have to wonder who the heck is sending me this information. You can't brand if people don't know from whom the email originated.

The sixth critical component of an online presence has to do with keeping your website current and up to date. Who trusts a website that has information from four years ago? There are a lot of zombie websites on the Internet, but nobody is home to keep them up to date. If you have outdated information, you might as well not post it.

The other challenge is that if you use an off-the-shelf program to update your website, you better be sure that it looks professional. Why have a beautifully designed website only to mess it up by making improper updates. The tool you use should be designed to work with your website;

otherwise, rely on professionals to make the changes for you. Why have a nice looking building on the outside, but then keep an unkept interior? You want to invite people into an experience that matches the external appearance.

The seventh critical component of an online presence is one that is most overlooked. When it comes to web hosting, your provider needs to be a partner in your business. After all, they are responsible for your brand's image. If a person tries to visit your website and they can't get to what they want right away, it leaves a negative impression. You must ensure that you have a provider that treats your brand like their own, instead of just another customer that's paying a few dollars per month. When you are promised resources, you better have those resources available to handle inquiries from your customers.

This next blog entry takes a reverse psychology approach. Since most people don't take good advice, I'll simply give them bad advice.

The BANNER BLOG

Don't Follow this Website Advice

http://www.bannerview.com/banner-blogs/mark-cenicola/story.bv?storyid=1295

Usually, when you tell someone not to do something, the first thing they do is just the opposite. If you found yourself intrigued by being told "no," then this headline probably captured your attention. This is not a trick, but just a little reverse psychology in action.

Below are 10 Great Website Tips to Being Successful Online:

1) Just have anyone build your website

A lot of people can design websites. Just ask your niece or nephew, or heck, just buy a book and learn to build your own website. Starting your business was easy so running it is probably just as easy. Who needs expensive professionals when you can get everything you need for next to nothing?

2) Never update your website

Who cares about fresh content? Post what you know at the time of building your website and live worry free knowing you never have to look at it again.

3) Make your contact information hard to find

After all, you are busy running your business. The last thing you want are customers calling, emailing or stopping by your place of business so that they can easily interrupt you. Plus, if they really want to do business with you, then they should at least spend some time looking for your contact information so definitely don't post your address or phone number in an easy to find place.

4) Never tell customers what you want them to do

If you haven't already learned to make your contact information hard to find, then you should certainly not tell your customers what you want them to do. "Call now," "email us today" or "click to buy" might make it so you actually have to do work. Plus, who wants to be seen like a pushy salesman?

5) Don't post anything informative

Your information is valuable so why would you want to give it away for free? People are always searching the Internet for free information. Let someone else waste their time giving it away.

6) Being found in the natural (organic) results on the search engines is overrated and a waste of time

Why bother trying to get listed on the search engines under the non paid section? You can just pay per click for keywords and get guaranteed placement in the sponsored results. How much could it cost anyway?

7) Working hard to be successful online is for suckers

Who wants to work hard, after all, isn't the Internet supposed to make our lives easier? Spending time making sure your content is up to date, your website can be found on the search engines and answering inquiries from bothersome customers is hard work left to suckers who have nothing better to do.

8) Don't get a website, just use social media

Facebook, Myspace, Twitter and Linked In are all free so why spend time and money on building a website? Since just about everyone is using them, I'll just let the social media sites own all my content and make them responsible to process my transactions. Besides, they are funded by millions of dollars in venture capital. It's not like any of them will fail or are likely to change their policies or be sold. My business reputation and brand is perfectly safe with them.

9) Don't bother building a strong online brand

With the recession huge brands that are many years old are going under or filing for bankruptcy. The consumer has lost trust in brands so building a consistent and strong online brand is totally last year. Don't waste time making sure your branding is consistent across your website, your email marketing messages or your blog. It doesn't matter if your brand gets lost throughout your online experience because no one cares about your brand anyway.

10) Many vendors are better than one single vendor as long as you're getting the cheapest price

You can pick a website from one vendor, email marketing services from another vendor, content management from another vendor, web hosting from yet another and they should all work together. When there's a problem, just start calling each of them and eventually one of them will help you fix it. If you are getting the cheapest price, why worry about saving time by dealing with a single company that can solve all your problems?

BONUS TIP: Prper speling isn't that impotant

Who reads websites anyway? With youtube, flash and cool graphics, no one is going to read what you put on your website. If you forget to spell check your website copy, no worries as it's not going to be read. Oh, and what's the deal with their, there and they're. Nobody cares if you aren't sure which "there" goes "where."

Okay, so the above list should have been called the 10 Worst Website Tips to Being Successful Online. However, since most people would rather give bad advice then listen to good advice you don't have to worry about people taking the above website advice or at the worst (best) case scenario they will take the advice and do just the opposite. If you want more advice like the above, DO NOT call BannerView.com at 702-312-9444 in Las Vegas or 888-221-8640 outside of Las Vegas and DO NOT get an Instant Website Quote here or DO NOT fill out a BannerView.com Website Proposal Request here.

Since you are actually reading this book, I'll provide you with some good advice; after all, if you've read this far, you are probably the kind of person who will follow it.

This next blog entry was inspired by a local political pundit. I asked him a question related to politics, but found that his answer actually applies to websites.

The BANNER BLOG

Political Advice from Jon Ralston Applies to Websites

http://www.bannerview.com/banner-blogs/mark-cenicola/story.bv?storyid=1289

While attending an International Association of Business Communicators (IABC) Las Vegas event, I had the opportunity to ask Jon Ralston a question as if he were to be on the other side of the table as a consultant to politicians as opposed to a political pundit. My question was about the advice he would give to a newbie entering politics. Of course, his first question was for what office am I running? While I don't have any immediate plans, it doesn't hurt to ask questions. Before even seriously considering running for office, all politicians could use his advice as it certainly made perfect sense. Based on the candidate pool, however, it seems not many have received such advice or if they have, taken it to heart. So what was his advice?

Jon Ralston's Advice for Newbies Entering Politics
* Have a reason to run for office (goal)
* Know the office for which you are running (research)
* Be honest (self explanatory)

How does this relate to conducting business online and why am I writing about politics? I'm certainly no political expert and furthermore no political pundit expert, but this advice, while it seems obvious, works for those businesses looking at entering the online world. So let's assume, that instead, I asked Jon Ralston what his advice would be for those newbies looking to enter the web. It might look something like this:

What Jon Ralston's Advice Might be for New Businesses Looking to Get Online
* Have a reason to get online (goal)
* Know what's online (research)
* Be honest (self explanatory)

Yes, okay, see how I've taken his words, applied it toward the subject of politics and spun it to be relevant to websites? Heck, I might make a good politician after all!

Actually, I don't call this "spin," but simply "leverage." If we leverage Jon's advice and simply apply it to other areas, whether it be politics or business, it is still quite relevant.

We meet with existing business owners, new business owners and prospective business owners all the time. However, many know that they need to get online with a website, but aren't quite sure why or what they will accomplish by doing so. Defining the goal is the first step before getting online and is the first advice we give. The next step is to research what competitors are doing to try to gain an advantage or at least be on par with them. Without those first two steps, success online is going to be hard to define, leave your competitors at an advantage and could possibly lead to other faux pas such as exaggeration and dishonesty to try to obtain results. We know that the majority of the time that simply leads to failure.

Most great advice has nothing to do with conducting business, whether online or off. However, when you see an opportunity to leverage good advice and use it in your own business, do it.

Since you obviously follow great advice, you are also probably the type of person who is actively involved in the process of building your online presence and a great brand for your business. Therefore, I'll share a blog entry with you that shows how your involvement leads to online success.

Customers that are Involved in the Web Design Process Fare the Best

http://www.bannerview.com/banner-blogs/mark-cenicola/story.bv?storyid=1294

We have customers across many different industries, different backgrounds, different experience levels and different views on how websites should be built. While the reason for hiring a company to help establish your online presence is to obtain the knowledge of an organization that's an expert at building websites, the companies that get heavily involved in the process seem to fare better than those that don't. That may seem obvious, but many companies hire Web design firms, pay them handsomely and say, "Just tell me when it's done." Then, when the company delivers the finished product, they wonder why it wasn't built the way they wanted it, why it doesn't portray what that company does and how it's going to bring them business.

At the other side of the spectrum you have companies that hire Web design firms, pay them handsomely, then wonder why they need to be involved in the process at all. Their thinking usually goes "Aren't we paying you to do the work?" Yes, you are, but those same Web design firms should be experts and have passion for designing websites. They aren't necessarily experts in your particular field, have the passion you have for your business or know the inner workings of your business. They need your help at producing a viable eBusiness solution.

Working together produces the best results

We've purposely designed our process to poke and prod the customer (yes it can be painful) for answers to questions, gather feedback and obtain active involvement. The customer doesn't necessarily have to be the business owner, but staff that's held accountable for the success and viability of your eBusiness initiatives. We want to build great websites that bring customers business and produce a nice return on their investment. That's why designing great websites is not easy! Such as it holds true with any successful business operation, it takes hard work from all parties involved, in this case the web design firm and the customer.

Successful customers remain customers

Most companies say customer service is their number one priority. Does customer service matter if you don't have customers to serve? Regardless of how great your customer service may be and no matter how great a product or service you provide, if your customers don't remain in business, they aren't going to remain customers!

So, when we push and push hard to get our customers involved and they try to rebut with those statements and questions above like "Just tell me when it's done, or "Aren't we paying you to do the work," we say, "If you didn't want to get involved in such a critical component of your operation like your website, does that also mean you don't want to be involved in keeping your company's doors open?

In nearly a decade of designing websites and complete eBusiness solutions, those customers that are actively involved in the process still remain our customers and have seen their businesses survive and thrive during even the most difficult of times.

So it's time to ask yourself the question, are you actively involved with your Website?

CHAPTER 10
THE SMALL BUSINESS COUNTER TERRORIST

*T*hink of the business you're in as counter-terrorism.

At BannerView.com want to be the big player, so we act like the big player in the web development market (at least in our market). We have "troops" on the ground attending various business-to-business networking functions. We have cars wrapped in company branding like a fleet of military vehicles. We regularly issue propaganda via email and blogging. And we spend more money than our competitors on advertising.

Yet, everywhere we turn, we are faced with competition. I call the competition "terrorists." Terrorists drive fear into people's minds. They act alone or in groups, and usually cause some sort of harm. In the examples to follow, the harm they cause is financial.

Most companies try to fight these terrorists at their own game by reducing pricing to be more "competitive." However, this strategy only causes a downward spiral of competition based upon price. It does nothing to bolster your brand. As the saying goes, "You may not always win the battle, but it's winning the war that matters."

Terrorist #1 - Amateurs

You would be surprised how many times when we ask, "Who's doing for your website?" we hear, "Oh my brother's girlfriend's son's cousin. He's a web designer."

What this answer really means is that some kid is good at surfing the Internet, so he's an assumed expert at designing websites. Well, like most people, if you can use a computer, type, and surf the Internet, you can probably figure out how to put two HTML codes together to create a web page.

You know the value a professional in your industry, whether you're a carpenter, business coach, attorney, or CPA. A professional brings to the table experience and expertise that an amateur can never bring. No matter where your service industry experience lies, you'll find yourself competing against amateurs, so why fight them on price?

The counter-terrorist response would be to politely let them know that you aren't scared by the amateur's low price, and subtly strike back by letting them know you'll check out their website. Then add them to your email marketing list, which serves as an educational tool that describes the advantages to using a professional for the services they need. In our case, that's a biweekly newsletter that provides tips and advice for conducting business online.

Amateurs are, and will continue to be, competitors that are nearly impossible to defeat. But they can be contained.

Education is the best weapon against them. Informed customers are not only better customers, but they also understand the difference in value between hiring amateurs and hiring professionals.

Don't bother competing head-on with amateurs. Let them have their day in the sun, and remember they will never be in a position to build a sustainable brand.

Terrorist #2 - Your Prospects' Staff

A lot of people will group web design with Information Technology, and assume it's a job for I.T. professionals, but I.T. professionals are there to make sure your computers work, your network is up, and that you can benefit from the Web's vast resources. I.T. professionals are probably not the best brand ambassadors.

I started my career as a network engineer for a technology consulting company on a city government contract. Back then, I was not the person to entrust something so important as communicating a brand to the world, especially after the company I worked for bounced a paycheck and gave me reason to venture into the business world full time. The best person for communicating a brand was the president of the company, who didn't have much of an I.T. background. While I could have struggled to put together a website, it was out of my area of expertise.

When we ask the question, "Who handles your website?" and get a response that it's handled by their staff "I.T. guy," we are very skeptical. We view the I.T. guy as another terrorist we must deal with.

In your profession, I'm sure you hear about how so-and-so on staff handles important issues, and you know they aren't the best qualified to handle it. No matter where your experience lies, you'll find yourself competing against in-house staff that prevent you from making sales.

Once again, the best weapon against in-house experts is education. Educating the customer about why it's important to bring in outside help for these kinds of projects. Your brand informs and educates, establishing you as an expert whose opinion should be listened to over that of in-house staff.

How many times do you find yourself more comfortable discussing important issues with complete strangers than your closest colleagues? It's simply easier to get an outside prospective that can be trusted. In the following raw entry, taken straight from my blog, I further hit home the point that in-house staff can be a source of terrorism against the success of your business.

The
BANNER
BLOG

Confusing Technical Expertise with Business Expertise Causes Bad Business Decisions

http://www.bannerview.com/banner-blogs/mark-cenicola/story.bv?storyid=1187

Many of our current and potential clients are at the size where they've hired the services of IT professionals, whether in-house or outsourced, to assist with the daily needs of using technology in business. Whether it's eradicating a computer virus, keeping the file server humming or implementing an information backup/disaster recovery strategy, these professionals are their jack-of-all-trades fire extinguishers. They are relied upon for their ability to swoop in and fix technical problems that are keeping organizations from getting the job done.

However, we see it time and time again that the reliance on fixing technical issues starts to turn into a reliance on these same experienced technicians to make business decisions.

For example, a company may tell their IT professionals that their computers are too slow and that their phone system can't handle their call volume and simply delegate the task to solve the problem. These IT professionals jump at the chance to obtain new technology and to put their skills to use so they start shopping around for vendors, compiling pricing and making decisions as to what to buy. This all may sound familiar and you may be asking yourself what's the problem with asking technology people to make technology decisions? Understandably, it seems simple enough for the IT professionals to just do their jobs by getting faster computers and buying a phone system that can handle increased call volume.

WARNING! Such thinking can be hazardous to your business' health!

In the end the IT professionals were happy because they did their job since the company got faster computers and a new phone system to handle a higher call volume. This caused the company to hire more employees to handle the increased call volume and order even more computers for those new employees. But wait! After all that time and money was spent, the executives were left wondering why more orders weren't being placed and why the new computers were still too slow. So the executives ended up consulting with an outside business expert (which cost even more money) to review the business' processes. The business expert determined the following reasons as to why the call volume and computers were slow:

* People weren't finding the answers to their questions on the Web site since it hadn't been updated recently and had to call in for support increasing the call volume.

* The online ordering process was confusing and users were either leaving the Web site or calling in to place orders, increasing the call volume.

* The Web site was hosted on the same server as the main file server. Since all computers in the office used that for file sharing purposes it slowed down all computers that were connected to it. In addition, it slowed down the Web site and made people call to place orders further increasing the call volume or simply frustrating users to the point of leaving the Web site altogether.

It just so happened that prior to the trouble with the computers and call volume, the IT professionals made the decision to keep the company's Web site in house even though it was not their forte. They figured it would be easy to maintain the Web site, would save money by not going with an outside vendor and ensure their job security since it provided additional responsibility.

Even though the above scenario is made up, this is a classic example of confusing technical expertise with business expertise. If the IT professionals would have had business expertise, their first questions might have been why are the computers too slow and why can't the phone system handle the call volume? Instead of turning over, what appeared to be a technical issue, the executives could have answered these simple questions by looking at the underlying business operations. From the beginning, if the Web site had been outsourced to an experienced company, new computers and a phone system wouldn't have needed to be purchased and additional employees wouldn't have had to be hired. In the end, the company would have saved so much more money by investing in their Web infrastructure which is a business decision, not a technical decision. Sales and profits would have increased and maybe the company could have ended up creating new higher paying jobs in production, sales & marketing.

This is why many larger companies hire a Chief Information Officer (CIO) so they can avoid such scenarios in the first place. A CIO is an executive level professional with leadership capabilities, business acumen and strategic planning experience. The CIO is generally responsible for processes and practices supporting the flow of information and it's quite common for CIOs to be appointed from the business side of the organization, especially if they have project management skills. Even though it's valuable for the CIO to have technical experience, they are not generally directly responsible for technology infrastructure. It's their job to hire the technical expertise to solve the organization's business needs. When it comes time to do business online it's wise to try to think like a CIO. Realize that technology is available, but make a business decision first to utilize that technology.

At BannerView.com we don't just make Web sites, you can count on our business expertise first and our technical expertise second to implement the most efficient Web based solutions.
It's true that technology can help make an organization's operations run more efficiently, generate more revenue and reduce overhead, but technology is only as good as the business decisions made to use that technology.

Related Article: BannerView.com Newsletter Volume #108 - Who's Behind the Development of Your Online Business?

Internal staff in a for profit business will eventually turnover; this means that those preventing a company from moving forward will leave. When those holes in a company's staff open, you will have the opportunity to win that business by positioning your brand properly.

Terrorist #3 - Big Businesses

As a small business owner, it's obvious that you are facing competition from the "big boys" in your industry. You may ask yourself how the heck you are going to compete with them. They have more money, more employees, more advertising dollars, more pricing power, and more of everything!

Big businesses use this power to infiltrate small business focused organizations such as chambers of commerce, and they garner a lot of political power in the process. Big businesses use their size to terrorize such organizations into driving customers into their nets.

The good news is that big businesses, in all their bigness, usually cast very big nets when they go fishing. A lot of small fish escape through those nets; the kind of small fish that can be mighty tasty for small business brands, and easier to digest.

Instead of trying to compete head-to-head with the big boys, here's your chance to be the counter-terrorist instead. By hanging around the swarm, you can find ways to cut small holes in the big-business nets, so small fish can escape right into your hands.

The best weapon of choice against big business is to bring your brand to the customer. Go where big businesses can't or don't go. Small networking events that are sometimes too small for big businesses to waste resources on are a great way to find those small fish. Get close to the customers that big businesses can't. Meeting face to face with potential customers is much more powerful than any other medium.

After an introduction, you can manage the follow up. By doing so, you create opportunity for your brand to become more personally connected to the potential customer. Once that personal connection is made, it's much harder for it to be broken by a big business.

A lot of times, small-business owners will say, "I'm going to think, act, and become BIG." So they start out with a giant net and cast it over a large audience without realizing that, should they catch the same size fish as big businesses, they will either crumble trying to fulfill all those big-fish orders (compromising brand goodwill) or lose all the tasty small fish in the process.

Look at what Apple has done with their physical Apple Retail Stores. They put "Geniuses" in the store with whom you can talk about your computer troubles. You can set an appointment to bring in your computer, and they'll work with you to fix your problems and answer all of your questions, free of charge! They brought their brand directly to the customer by casting a much smaller and personal net.

This created trouble for small Apple related vendors since the company was going to where they lived. However, in the process, Apple created the reverse problem. What used to be a few minute wait to get in front of an Apple genius can now take hours. That net caught so many fish, it's now hard for them to manage all of the appointment requests. In turn, it's again created opportunities for small vendors to handle support for those customers who don't want to wait. If you are one of those vendors, cast your net just outside an Apple Store and see what happens!

Positioning your brand as one of differentiation from the large business brands is a key to success against such foes. Your resources are simply too small to take them head on. Build upon added value, specifically the close personal relationships you can build with your customers. When people feel like they have an "in" with people who can effect change, it sends a powerful brand message that you are on their side while casting the big businesses as greedy monopolies.

Terrorist #4 - Other Small Businesses

Don't think I'll leave out the biggest terrorizer of small businesses: YOU!

Since small businesses certainly can't be everywhere, neither can you. That leaves the door open for other small businesses to get your customers. However, at the same time, small businesses get caught up in the day-to-day operations, so it's hard for them to think beyond their immediate business prospects and build long term brand exposure programs.

For example, how persistent are small-business owners in selling you their products and services if you don't display an immediate interest? Not very. They are busy trying to find the next hot lead and convert that lead into an immediate customer. Why waste time on prospects who don't show an immediate interest?

This is where a persistent form of follow-up brand exposure comes into play. The best weapon of choice against other small businesses and other terrorist groups are actually multiple weapons that work together: **blogging, email marketing, networking, authoring, public speaking, video, websites, and social media.** Doing these things right can build lasting brand recognition and personal attachment to your brand. That's why a chapter of this book was devoted to each topic.

So in reality, I'm not the CEO of a company that builds great websites that earn recognition, respect and rewards. I'm a counter-terrorist who is constantly engaged in a counter-insurgency against amateurs, our prospects' staff, big businesses, and other small businesses. Like any counter-insurgency, the process is long and expensive. That's why it's more important than ever to find long-lasting, low-cost means of waging the never-ending brand-building battle on your own terms to effectively out maneuver your competition.

CHAPTER 11

GET THE TOOLS TO BUILD YOUR BRAND: SHAMELESS SELF PROMOTION

*T*his chapter is mostly self serving (as if the other chapters weren't). If you aren't interested in obtaining the tools to make your brand a success, feel free to stop reading now. Actually, while this may be a sales pitch, it's important that we share the reasons why we've come to this point.

In our business, we realized that small businesses can never really compete on the same level as large businesses. Even with all the tools at low prices, you can see that it's not enough to simply have the tools; it's important to have the right tools that work together, and know how to properly use all of those tools.

We realized that there needed to be a solution that brought all the right tools together and made it affordable to the small business owner. We had to make it possible for owners to build a great brand because some of those brands will eventually become household names. There's no better satisfaction than being associated with, and having been a part of, a success that is even greater than the success of our own brand.

The following blog entry speaks to is issue. We created an innovative way to get small businesses all the tools they need to build a successful brand online at an obtainable price. Now you can enjoy the success that was previously only available to large businesses.

The BANNER BLOG

Bringing Professional Web Sites to the Masses

http://www.bannerview.com/banner-blogs/mark-cenicola/story.bv?storyid=1200

After more than 9 years of developing professionally designed and highly custom Web sites for hundreds of customers, not much has changed in terms of affordability, even with all the advancements in technology. Customers are left to choose between a professional Web development agency usually costing thousands of dollars for a custom Web design or consider doing it on their own. Obviously, it's night and day between the two choices. Rare, but sometimes attainable, a company could find a freelance or moonlighting Web designer to develop a professional site for an affordable fee. I say "rare" because it's difficult to find a single designer who has both a business and design background in order to deliver a product that meets the needs of the business requesting such services. In addition, they are usually very busy with their day jobs or your project takes a back seat when more pressing items are on their agenda.

Just with any construction project, whether building hotel casinos or online businesses, it takes a team of experienced people to build a quality product and there's certainly a cost for that experience. Over the years, many customers have come to us with the perception that doing business on the Internet is inexpensive, sometimes even free! By the time a customer gets a custom designed Web site, has the site optimized for search engines, adds promotional tools such as email marketing or business tools like e-commerce and sets aside money for ongoing Web hosting & maintenance, the costs are several thousands of dollars. While doing business online is certainly a fraction of the cost of opening a physical location, as you can see, there are costs involved none-the-less. A number of customers with good ideas and viable business plans have been unable or unwilling to invest the time and money into their online presence to help grow their businesses. While we wish this wasn't true, the reality is that it takes money to make money. Those that have it, can use it to expand their businesses, while those that don't must continue to grow slowly and struggle to survive. We certainly have experienced the ups and downs of being an entrepreneurial company, needing money when there isn't, making do with what we have and taking the calculated risks when needed.

As a company that relies on its Web operations as the very core of its existence, we understand first hand how important it is for any company to have a great online presence to help drive growth, reduce costs and save time. Unfortunately, you can't call your banker and ask for a loan to develop a great Web site (if you can you'll have to introduce me to your banker). It requires the use of working capital from your marketing budget or other type of funds to get your site off the ground even though it may be a core asset of your company's operation. A Web site has never appropriately been treated as a financeable asset and even if it were, in this credit climate, good luck getting financed. A great many number of companies would have had much better Web presences should there had been a common financing mechanism for Web development like there is with traditional development.

Being that the financing side of the equation has always been an issue for companies when it comes to getting everything they need to be successful online, we set out nearly two years ago to find a solution to this problem. After several stumbles and dead ends while trying to bring in financial partners to assist, we made the decision late last year that we'll do something about it ourselves. Just as the credit climate became a disaster, our decision turned out to be one of the best decisions would could have possibly made.

That's why yesterday, we officially announced our Web site Solution Builder that gives companies of all sizes the opportunity to get a custom developed Web solution with all the features to help make them successful at an affordable monthly price. Unlike any other solution we've researched, the Solution Builder makes it as easy as picking out the options on your favorite car, but unlike with traditional financing, there are no credit checks, interest fees or haggling because your 1, 2 or 3 year pricing options are clearly displayed. In addition, that monthly price you see is inclusive of your Web hosting & maintenance! While we'll continue to offer our solutions with traditional up front payment options, our new lease options make it affordable for just about any business to get the tools they need to be successful online. Now, instead of having to drop thousands of dollars up front on your Web presence, you can use that money to help market your business or for operating purposes.

As I've hinted over the last couple of months, we feel this is the solution needed, especially in this market, to help businesses, not only do business online, but do business online right!

We understand that an educated customer is a better customer. That's why, even though we may be hitting you with a sales pitch, we still provide great value behind the pitch.

We've all had our fair share of selling customers who turn out to not be the brand ambassadors we'd hope, so our focus is getting customers to buy when they are ready. We want buyers—not people to whom we can sell. We want our brand associated with small businesses that don't need to be sold, but small businesses that realize the solutions we have are the best opportunity for them to build successful brands.

This next blog entry speaks to exactly that point.

If They're Ready to Buy, You're Ready to Sell

http://www.bannerview.com/banner-blogs/mark-cenicola/story.bv?storyid=1185

I've learned some very valuable lessons over the years about the sales process. The biggest lesson I've learned, and it really hits home in our current economic market, is that no matter how great a sales person you are or how great the product or service offering you have is, you won't get people to buy your products or services until they are ready to buy. So instead of trying to figure out the best sales process, best pricing strategies or perfect marketing verbiage, focus on the people that are ready to buy. This may seem obvious, but it's often overlooked. I've found myself following up with potential customers after we've provided all the information, product details and solution overviews that they've requested to still have them not buy. I'm not saying that you should ignore following up with potential customers, but if you've provided them with all the information necessary for them to make a buying decision and they haven't moved forward, they're simply not ready to buy. Calling them every week or emailing them constantly isn't going to make them anymore ready.

So what do you do? You must know the primary motivations behind why people buy:
* **Need** - People need the product or service you are selling.

* **Convenience -** It's convenient for people to buy the product or service you are selling.

* **Scarcity -** What you are selling is rare.

* **Prestige -** People buy to make themselves feel more important than others.

* **Emotional Fulfillment -** People buy to substitute a void in their lives.

* **Lower price/affordability -** People wanted your product or service before and now they can afford it.

* **Value -** What you are selling, at the price you are selling it, exceeds the perceived value of the buyer.

* **Brand -** People associate your product or service with being the best, most recognizable or best solution for their needs based on name alone.

* **Innovation -** Your product or service is the latest and greatest.

* **Compulsion -** A person buys because they have no other choice in the matter.

* **Ego -** People make purchases to impress others and stroke their ego.

* **Association -** People make purchases from people they want to be associated with.

* **Peer Pressure -** Everyone else is doing it so why don't you.

* **Reciprocity or Guilt -** People buy because someone bought something for them or because they've been guilted into buying.

* **Empathy** - Someone buys from you because they care for you.

* **Addiction** - People buy because it's hard to say no.

Before you make that next follow up phone call to a potential customer, ask yourself, is what you are going to say to them going to change their buying motivation? For example, if a customer didn't need your product or service before, do you have a reason they will need it now? Has the value of your offering increased significantly to make the customer more responsive to purchasing? Have you lowered the price or made your product more affordable? Have you made it more convenient to purchase from you or obtain your products and services more easily? Have you found out that one of their competitors is doing what you are proposing and they would be missing out if they didn't do so as well?

When customers are ready to buy they will certainly get back in touch with you. I know it's tough in this economy, but patience is a virtue. You are just going to have to wait until they are ready. If you can't afford to wait, then you must go back and revise your offering to fit within the motivations behind why people buy. The best customers are those that are ready to buy your product or service on their terms as they are less likely to have buyer's remorse, will usually be long term customers and will speak goodwill of your company. Take a look at the sales process from a different perspective. If someone is ready to buy, only then are you ready to sell.

Now that you've been warmed up to the fact that I really don't want to sell you something, it's time to give you a reason to buy. This next blog entry speaks to our unique selling proposition when it comes to getting you all the tools you need to be successful online at an obtainable price.

The BANNER BLOG

Website Leasing: What is it, How Does it Work and How Can I Benefit?

http://www.bannerview.com/banner-blogs/mark-cenicola/story.bv?storyid=1223

One of the biggest challenges both small and medium businesses have, when it comes to getting a custom website developed inclusive of things like e-commerce, email marketing, website optimization, website promotion plus web hosting and maintenance, is the affordability factor. Getting a professional website when including all the parts necessary for a successful online presence can be quite expensive.

When the cost to do things right is out of the reach for small and medium businesses, they do one of three things:

* Skimp on the important things that are critical to driving business to a website such as email marketing systems and website optimization.

* Turn to companies offering cheap, generic looking websites thrown together without much thought for how the

site is supposed to consistently deliver the brand's message, generate leads or process transactions online in a secure and efficient manner.

* Do nothing.

So what's a business to do when they obviously need to be able to compete effectively in the digital age without exhausting much needed working capital? The answer may be to **lease your website.**

Leasing, when it comes to websites, isn't an option that's often considered because:
* Leasing simply doesn't come to mind. Traditionally, when people think about leasing they think about leasing tangible products like cars, computer hardware and office equipment.

* Many banks won't finance an intangible asset. Websites, much like software, are seen as intangible assets for which most banks will refuse to provide lease financing because it's not something that they can easily sell should you not be able to pay.

* Financing is difficult in this market. Even if financing was available on an intangible asset like a website, getting approved in this economy requires great credit on behalf of the borrower.

* Confusing terms can cost you. Many leases are written to favor the lender, not the borrower which means that you could find yourself paying outrageous fees to purchase the asset outright, get stuck with high interest rates or fall under terms that automatically renew themselves for many more years.

Knowing that companies of all sizes are looking for ways to compete online, we developed a website lease program to make obtaining a successful online presence, at an affordable monthly price, a possibility. We'll take a look at what it is, how it works and how your business can benefit.

What is Website Leasing?

Leasing a website is very similar to paying a monthly subscription fee. Instead of paying a lump sum for your website up front or incurring setup fees, you pay a much lower monthly fee for a set period of time. BannerView.com offers website leases for 1, 2 & 3 year terms with the longer the lease term the lower the price.

How Does Website Leasing Work?

Using BannerView.com's Solution Builder, you select your website package and add any additional options that are right for your business to help build, promote and maintain a successful online presence. The monthly fee is clearly displayed based upon the options chosen for your website for each of the 1, 2 & 3 year lease terms. What it would cost you to purchase the website as opposed to leasing it is displayed for those wanting to buy it outright, but also serves as a comparison to see what you would have had to invest up front to get the same solution. You can also see the cancellation terms and buyout options if you choose to own your website after the lease term is complete. There are no credit checks and getting started only requires the first and last month's lease payment.

How Can My Business Benefit from Leasing a Website?

Getting all the tools needed to build, promote and maintain a successful online presence can be a significant investment. BannerView.com's website lease option provides the ability for you to now afford those tools that will give you the best chance at success on the Internet while preserving your working capital. The website lease is a great option even if you're use to owning and paying for everything up front. After the lease period is over you have the option to buyout the website, continue month to month or walk away. Since no banks are involved and no credit checks are required, there's really not much stopping you from getting into a website lease!

Never resting on our laurels, we continue to focus on the core message of our brand. When we decided upon our core brand message, it was important to jettison those things that didn't fit within that message, as the next blog entry discusses.

The
BANNER
BLOG

Focus on What You Do Best & Sell the Rest

http://www.bannerview.com/banner-blogs/mark-cenicola/story.bv?storyid=1221

We recently completed the sale of our DataCities.com operations which we only just purchased in November 2007. The question of why we bought it in the first place and why we sold it so quickly are great questions which have to do with our company's focus on what we do best.

Q) Why did we buy DataCities.com in the first place?

A) We originally purchased DataCities.com to gain access to Lpanel, a Web hosting and billing automation software suite that integrates directly with cPanel. Many independent Web hosting companies rely on cPanel to power their Web hosting operations such as setting up accounts, provisioning resources and most things needed to handle the technical details of running a Web hosting operation. Lpanel handles all of the administrative aspects and helps those cPanel Web hosts with the ability to manage both their billing and support operations and provide customer service. Together they are like peanut butter & jelly.

We figured purchasing DataCities.com along with our acquisition of Lpanel was a great way for us to test the Lpanel software in a real world situation - on a Web host that provides services to real customers with real support, technical and billing needs. And it was! We were able to release over 50 updates that fixed a number of critical and minor bugs and added new features for Lpanel in just over 1 year. Of course, the existing Lpanel customer base also helped tremendously in helping us launch new versions, but having first hand experience significantly accelerated this process.

Q) If DataCities.com was so successful in helping to develop Lpanel, why sell it now?

A) We are now in a position that the current development tree of Lpanel is stable and doesn't need so many frequent updates. Most updates no longer require real world situations to test their effectiveness, however, a number of our customers have been more than helpful when it comes to testing bugs in real world situations.

As a separate operating unit, DataCities.com was focused on cPanel reseller Web site hosting which is not in line with our primary operations. The sale gives us the opportunity to continue our focus on our BannerView.com suite of products (like Lpanel) and services, specifically our new Web site leasing options.

Q) Did this sale have any effect on BannerView.com employees or customers?

A) Since DataCities.com was operated as a completely separate division within our company there was no effect on BannerView.com's primary operations. No BannerView.com customers were included in the sale and no current employees were transferred to the new owners of DataCities.com. We'll continue to support and maintain Lpanel as a BannerView.com product.

With us retaining the Lpanel software and technology, we can continue to use the best parts of what we initially bought while freeing up operating cash and providing the new owners an asset in great shape that fits nicely within their focus.

Specific aspects of the Lpanel technological framework have been incorporated into other BannerView.com products such as BannerMailer and plays a critical role in our newest innovations when it comes to Leasing your Web site at an affordable price while keeping money in your pocket.

Q) What can be learned from our experience?

A) Take a lesson from our playbook and evaluate your current operations. Decide which pieces still make strategic sense, continue to focus on those areas and sell the rest! You'll free up valuable resources that you can really use to survive and thrive in this economy. And if you need help figuring out where to focus, what to sell or help selling it, a great resource is the Turnaround Management Association. My involvement with the Nevada Chapter of TMA, has provided the insight needed to make strategic decisions in difficult times and become associated with the individuals that can make things happen.

Relying on only one area of your online presence will lead to a false sense of security. It's necessary to do them all together to get the greatest benefit, which will lead to building your own banner brand.

APPENDIX

Here is a quick reference guide to a list of all of the blog articles mentioned in this book along with the URLs so you can view them in their original format. They are listed here in the same order as they do in the book.

The Age Old Debate - PC Versus Mac (Price versus Cost)

www.bannerview.com/banner-blogs/mark-cenicola/story.bv?storyid=1214

Let the Tours Begin

www.bannerview.com/banner-blogs/mark-cenicola/story.bv?storyid=1290

Intermingle Current Customers w/Potential New Customers | The BannerEdutainment Tours Roll On!

www.bannerview.com/banner-blogs/mark-cenicola/story.bv?storyid=1298

Communication is Key in Customer Service

www.bannerview.com/banner-blogs/mark-cenicola/story.bv?storyid=1227

It's Not Change, It's Progress

www.bannerview.com/banner-blogs/mark-cenicola/story.bv?storyid=1196

Small Improvements Can Make a Big Difference | Moving Forward Despite a Good and Bad Economy

www.bannerview.com/banner-blogs/mark-cenicola/story.bv?storyid=1300

Combined Online and Offline Marketing Efforts Lead to Marketing Campaign Success

/www.bannerview.com/banner-blogs/mark-cenicola/story.bv?storyid=1211

What's Your Google Results Page Look Like?

www.bannerview.com/banner-blogs/mark-cenicola/story.bv?storyid=1323

Web 2.0 This, Social Media That, Blog Here, Blog There, Blog Everywhere

www.bannerview.com/banner-blogs/mark-cenicola/story.bv?storyid=1156

If you Blog It They Will Come: Success in Professional Blogging

www.bannerview.com/banner-blogs/mark-cenicola/story.bv?storyid=1168

Professional Blogging to Generate Business: The Power of PR & Branding

www.bannerview.com/banner-blogs/mark-cenicola/story.bv?storyid=1174

Blogging for T-Shirts

www.bannerview.com/banner-blogs/mark-cenicola/story.bv?storyid=1189

Blogging Overload is Delivering Website Traffic

www.bannerview.com/banner-blogs/mark-cenicola/story.bv?storyid=1248

Tips to Finding Things to Write About for Your Blog | The BV Ride A Part of History?

www.bannerview.com/banner-blogs/mark-cenicola/story.bv?storyid=1280

The Ethics, Laws and Tips to Success with Email Marketing

www.bannerview.com/banner-blogs/mark-cenicola/story.bv?storyid=1191

Hitting Them from Every Angle Makes Closing the Deal Easier

www.bannerview.com/banner-blogs/mark-cenicola/story.bv?storyid=1286

Don't Let Your Email Marketing Suck

www.bannerview.com/banner-blogs/mark-cenicola/story.bv?storyid=1337

Your Website Can Lead to Engagement (Literally)

www.bannerview.com/banner-blogs/mark-cenicola/story.bv?storyid=1315

Looking the Part Plays a Vital Role in Developing a Successful Online Brand

www.bannerview.com/banner-blogs/mark-cenicola/story.bv?storyid=1383

Lining up the Speaking Gigs

www.bannerview.com/banner-blogs/mark-cenicola/story.bv?storyid=1242

Speakers, Trainers & Coaches Geared up to Bust Out!

www.bannerview.com/banner-blogs/mark-cenicola/story.bv?storyid=1320

Educating the Windy City at the Chicago Bust Out Boot Camp

www.bannerview.com/banner-blogs/mark-cenicola/story.bv?storyid=1303

Not Familiar with Twitter Yet? It's a Billion Dollar Company!

www.bannerview.com/banner-blogs/mark-cenicola/story.bv?storyid=1218

Web 2.0 - Are You Giving up too Much Control to It?

www.bannerview.com/banner-blogs/mark-cenicola/story.bv?storyid=1237

Let the Social Media Shakeout Begin

www.bannerview.com/banner-blogs/mark-cenicola/story.bv?storyid=1249

Strategic Social Media Planning

www.bannerview.com/banner-blogs/mark-cenicola/story.bv?storyid=1317

A Complex Formula for Finding Me Where I Work!

www.bannerview.com/banner-blogs/mark-cenicola/story.bv?storyid=1283

The Power of Web Video and Planning for Its Success

www.bannerview.com/banner-blogs/mark-cenicola/story.bv?storyid=1328

Your Website is Not Properly Coded and It's Costing you Business!

www.bannerview.com/banner-blogs/mark-cenicola/story.bv?storyid=1334

Designing Great Websites is NOT Easy

www.bannerview.com/banner-blogs/mark-cenicola/story.bv?storyid=1243

Don't Follow this Website Advice

www.bannerview.com/banner-blogs/mark-cenicola/story.bv?storyid=1295

Political Advice from Jon Ralston Applies to Websites

www.bannerview.com/banner-blogs/mark-cenicola/story.bv?storyid=1289

Customers that are Involved in the Web Design Process Fare the Best

www.bannerview.com/banner-blogs/mark-cenicola/story.bv?storyid=1294

Confusing Technical Expertise with Business Expertise Causes Bad Business Decisions

www.bannerview.com/banner-blogs/mark-cenicola/story.bv?storyid=1187

Bringing Professional Web Sites to the Masses

www.bannerview.com/banner-blogs/mark-cenicola/story.bv?storyid=1200

If They're Ready to Buy, You're Ready to Sell

www.bannerview.com/banner-blogs/mark-cenicola/story.bv?storyid=1185

Focus on What You Do Best & Sell the Rest

www.bannerview.com/banner-blogs/mark-cenicola/story.bv?storyid=1221

www.ingramcontent.com/pod-product-compliance
Lightning Source LLC
Chambersburg PA
CBHW022042210326
41458CB00080B/6592/J